Shifting Time

Shifting Time
Social Policy and the Future of Work

Armine Yalnizyan,
T. Ran Ide and
Arthur J. Cordell

Introduction by Jamie Swift

between the lines

Published by:
Between The Lines
720 Bathurst Street, Suite 404
Toronto, Ontario M5S 2R4

Cover design by Counterpunch

Printed in Canada
"Securing Society: Creating Canadian Social Policy" is published in co-
operation with the Canadian Centre for Policy Alternatives and the
Social Planning Council of Metropolitan Toronto. It was originally
published in pamphlet form as "Defining Social Security, Defining
Ourselves: Why We Need to Change Our Thinking Before It's Too Late."

Between The Lines gratefully acknowledges the financial assistance of
the Canada Council, the Ontario Arts Council, and the Ministry of
Canadian Heritage.

Canadian Cataloguing in Publication Data

Yalnizyan, Armine
Shifting Time

ISBN 0-921284-90-X (bound)   ISBN 0-921284-91-8 (pbk.)

1. Technological innovations—Social aspects — Canada. 2. Labor
supply —Effect of technological innovations on — Canada.
3. Canada —Social policy. 4. Wealth—Canada. I. Ide, Ran. II. Cordell,
Arthur J. III. Title.

T14.5.Y35   1994   303.48'3'0971   C94-931201-0

Contents

About the Contributors

Armine Yalnizyan is one of Canada's leading labour market analysts. As a part-time worker, she is the Program Director at the Social Planning Council of Metropolitan Toronto. As the full-time mother of three pre-schoolers, her research on labour market trends is rooted in personal experience. A frequent media commentator, Ms. Yalnizyan works with community-based groups on economic literacy and civil rights issues.

Educator and broadcast executive **T. Ran Ide** founded TVOntario. He spent twenty years as a teacher, principal, and school superintendent in Thunder Bay before becoming the first chairman and CEO of Ontario's public television network. He has been acting vice-president of and special adviser to the CBC and a member of the Science Council of Canada and holds honourary degrees from Queen's University and the University of Waterloo.

Arthur J. Cordell is a prominent thinker on the social implications of information technology. At the Science Council of Canada, he produced some of this country's most innovative policy ideas on the conserver society and the role of technology in our collective future. Dr. Cordell holds a PhD in economics from Cornell University.

Jamie Swift is a writer and the author of *Cut and Run: The Assault on Canada's Forests*, *Odd Man Out: The Life and Times of Eric Kierans*, and (co-author) *Getting Started on Social Analysis in Canada*.

Introduction

Jamie Swift

There was a story on the radio news this past Christmas about the efforts of charitable Calgarians who were doing their best to brighten the bleak winter prospects of a group of citizens less fortunate than themselves. The volunteers were helping to fill Christmas hampers with small gifts and food. It was the type of story you come to expect in the slow news days just before Christmas, something to remind us of the spirit of the season. The odd thing about this news item was that the volunteers were not packing toys and turkeys for welfare mothers or old men in rooming houses. The care packages were intended for a new group of impoverished people—employed hospital workers. These people at the low end of the health-care pay scale were having their paycheques reduced along with their access to hours as the government slashed its health-care budget in an attempt to completely eliminate the provincial deficit.

The report from Calgary was followed by one from Tokyo. The Japanese press had noticed a new phenomenon, the underemployed office worker. The surplus employees were quickly labelled "the tribe that sits by the window." These were men who were part of Japan's famous core workforce, the group that enjoys lifetime job security. Yet between one and two million of them were now being paid to sit and gaze out the window because there was little or nothing for them to do.

The news from Tokyo went on to point out that Japan's low unemployment rate of 2.5 per cent hid substantial underemployment and that the commitment of Japanese business to employment security in return for employee loyalty was fast eroding. Firms such as Nissan and Nippon Steel have announced mass layoffs to be phased in during the next few years. Others, from machine-tool makers to producers of consumer electronics, had lowered the retirement age to cut their workforces. This gave rise to another expression, this one coined by older Japanese women to describe husbands who suddenly found themselves at home with nothing to do: "fallen leaves."

These short reports from two of the wealthiest cities in two of the world's wealthiest countries distilled the massive challenges facing advanced capitalism. Many people are working longer hours than ever before, but unemployment and underemployment are steadily rising and seem unlikely to go away. More people are becoming dependent on the welfare state (or, as it has become metaphorically known, the "social safety net") at a time of high deficits and the ostensible necessity to cut or "modernize" the welfare state. Indeed, by 1994 it had become clear that the welfare state had reached a watershed in its history, with everyone from politicians to poor people's advocates deep in discussion of what is to be done in the face of a polarized, good jobs/bad jobs economy.

No one should doubt capitalism's ability to not only produce challenges but also take advantage of them. Growth sectors include the private security business and, of course, the home computer business. In a 1993 advertising insert entitled "Toward 2000: A Progress Report," the Xerox corporation unblinkingly— indeed faithfully—reproduced the prediction of media trend-spotter Faith Popcorn. The Xerox version of the millennial home is more than the electronic cottage, and at the same time is chilling and enticing to marketers of the much-discussed electronic highway:

> Home . . . will be a fortress, a workplace and an entertainment centre. Sophisticated security devices will guard most homes from crime. Indeed, security will become one of the fastest-growing businesses of the 21st century, especially if society becomes more polarized between the wealthy, educated "haves" and the illiterate, poor sociopathological "have-nots."[1]

The market-based system has gone through lean and mean phases in the past. But a key difference this time around (as this book shows) is that there is no upstart sector to take up the slack. The prevailing response of more growth and bigger markets, with particular hopes pinned on the Pacific Rim, is the road to environmental disaster. How to reconcile our capacity to produce a superabundance of goods and services with the obvious fact that growing numbers of people cannot afford basic goods and services? And how to deal with the stunning paradox of governments groping frantically for ways to deal with social and economic policy or to restore "growth," while at the same time searching for new places to bury the shells of microwave ovens that didn't cook fast enough and the sour toxins that are the byproducts of industrial production?

The poverty-in-wealth phenomenon, long apparent on a world scale, is being noticed once again in Canada. Politicians and policy-makers are scrambling for solutions, the most popular of which seem to revolve around cutting costs and "retooling" or "modernizing" social programs. Ottawa looks to Fredericton, where the provincial government of Frank McKenna is seen to be in the vanguard of reform. The approach transcends party lines.

The trend had become clear by 1992, well before the Conservatives had been replaced by the Chrtien Liberals, a group more easily identified with the New Brunswick Liberals. By March of that year New Brunswick had renamed some old ministries, and its Department of Advanced Education and Labour had got together with the Department of Income Assistance to produce a proposal with the catchy title "NB Works." The idea was to retrain jobless welfare recipients, providing them with new skills for the new jobs of a post-industrial economy whose main pole of job growth would be the service sector that has supplanted manufacturing, the nineteenth century's uptake sector. This would relieve government of the burden of supporting them.

By year's end it had become obvious that Ottawa had adopted the same approach. The Health and Welfare department (subsequently and symbolically renamed the Department of Human Resources) produced a paper on the "new orientations" of Canadian social policy that was tabled at a meeting of the Organization for Economic Co-operation and Development (OECD) in Paris in December. Like the NB Works proposal that Ottawa had embraced with a huge subsidy, the federal government was looking at a supply-side approach to the labour market that emphasized "breaking the spiral of dependency" through "self-sufficiency," "greater individual responsibility," and, of course, training to meet the demands of "global competition."[2] This is all part of what Armine Yalnizyan sees as "the new war of words."

It was a war that the new Liberal human resources minister, Lloyd Axworthy, leapt into when he announced a major 1994 review of welfare state programs. The government was concerned about the three million people "frozen into perpetual poverty" and was not out to "slash and trash" but to "redesign." Axworthy, a savvy politician aware of the weight of words, denied that redesign was a "codeword" for cutting spending.[3]

The war of words has replaced any notion of a war on

poverty. Indeed, Canada's new-look social policies are aimed at adapting lower-income Canadians (of whom there are increasing numbers) to their future as part of what John Kenneth Galbraith has called a "functional underclass."[4] This is a group that can function as a necessary underpinning of capitalism in the post-industrial age: that is, a class of workers handcuffed to a changing labour market that is producing a lot of poorly paid, part-time jobs; a class of workers whose expectations have been ratcheted down to the point that getting thirty hours of work at $8.50 an hour may start to seem like a real job — or perhaps the only job they can really aspire to. Some 18 per cent of the U.S. workforce is now made up of people who put in forty hours for poverty-level wages, a fact with enormous implications for Canadians in the era of free trade and a continental labour market.[5] According to Ran Ide and Arthur Cordell, "the rich do very well" in a business-as-usual scenario in which non-standard employment, shrinking social services, and private security systems not only co-exist but in fact are also interdependent.

This is the background for this book. Armine Yalnizyan, Ran Ide, and Arthur Cordell have conducted an examination of the *real* world of work, wealth, technology, and a Canada-in-transition. Their analysis is unencumbered by the newly fashionable talk of social welfare reform whose main message is that Canada is no longer wealthy enough to aspire to old-fashioned notions such as equality and fairness. This discussion is important for two reasons.

Firstly, it questions the values and ethics of a society and economy based on the maximization of wealth and the worship of technological progress. Self-sufficiency, self-reliance, and ending the cycle of poverty and dependence sound like fine goals. But, as Yalnizyan points out, the use of such discourse in the present context merely masks a nineteenth-century message designed to prepare us for the twenty-first, "with the poor being told they are responsible for themselves and the rich being assured they are responsible for no one but themselves." Similarly, Ide and Cordell question the technological imperative and the accompanying "secular religion of economic growth at whatever cost," including unsustainable costs to the environment. Instead, they posit a future free from today's idolatry of market ideology.

The contributors to this book make it clear that the kind of full employment that seemed attainable in the first twenty-five years after World War II is no longer possible. What's needed is a redistribution of work and wealth, of time and money. It is time to rethink the almost automatic equation of work with employment and in this way seek a real solution to what that humorous old Tory, Stephen Leacock, labelled "the unsolved riddle of social justice."

In 1920 Leacock, whose first career was as a respected McGill University political economist, marvelled at the co-existence of poverty and wealth. "The tattered outcast dozes on his bench" while the rich young "puppy" seeks to amuse himself in a state of "melancholy monotony." How could this be, Leacock wondered, when there was more than enough to satisfy our basic needs?

> The hours of labor are too long. The world has been caught in the wheels of its own machinery which will not stop. With each advance in invention and mechanical power it works harder still. New and feverish desires for luxuries replace each other as older wants are satisfied. The nerves of our industrial civilization are worn thin with the rattle of its own machinery. The industrial world is restless, over-strained and quarrelsome. It seethes with furious discontent, and looks about it eagerly for a fight. It needs a rest. It should be sent, as nerve patients are, to the seaside or the quiet of the hills.[6]

If Canada was overstrained and quarrelsome in 1920, how would Leacock react to his country in the 1990s, when the hum of the computer has replaced the rattle of machinery and public faith in politicians and government has reached an all-time low?

This brings up the second crucial point raised by the authors of *Shifting Time*. Both contributions show that we have to rethink the dominant approaches to dealing with the wrenching changes wrought by globalization, automation, and the apparent end of employment—at least as we have known it. They emphasize the need for new *ideas*. As Yalnizyan puts it, we need more than a meaningless and ultimately useless war of words. Canada needs a "new war of ideas" about the need for redistribution, because the labour market is increasingly incapable of producing anything but what Ide and Cordell predict is "a two-tier society where the poor and the rich will see each other across a great unbridgeable divide."

The authors present us with solid evidence (statistics, policy analysis, case after case of the new realities of production and distribution) of the bankruptcy of old ideas about progress. The "learnfare" approach to social policy of the "NB Works" variety promises only to institutionalize the great divide between good jobs and bad jobs, training people for good jobs they will never get or bad jobs into which they can be slotted, most likely on the precarious basis that characterizes a labour market offering part-time, contract, or temporary positions.

The alternative advocated by Ide and Cordell hinges on a new approach to productivity. Why not institute a "productivity tax" on the new computer and information technologies that are replacing so many jobs? This idea is as old as Luddism. In fact, when the "croppers" in their hundreds stormed and burned new mills at the very end of the eighteenth century, "proposals were in the air for the gradual introduction of the machinery, with alternative employment found for displaced men, or by a tax of 6d per yard upon cloth dressed by the machinery, to be used as a fund for the unemployed seeking work."[7]

Taxing productivity? The very notion seems unimaginable during a time when we are being so often told that the forces propelling the world in its current direction are inevitable. But it has a certain elegant simplicity, this idea that when society in general becomes more productive we should all benefit. Politics too often lacks a sense that the unimaginable is possible. Pragmatically, the notion of a productivity tax is particularly intriguing at a time when a dwindling layer of middle-income Canadians seems increasingly unwilling and unable to support public provision through higher personal taxes.

Yalnizyan argues convincingly that the corporate support for public provision has sharply declined over the past forty years as the tax burden has shifted from business to individuals. Indeed, the need for new ideas about distribution comes up repeatedly in what follows. She recognizes that international business is well positioned to avoid taxes by fleeing — the postmodern version of the capital strike. And she makes a simple-sounding proposal, that Canadians and the citizens of other lands demand the right to produce what they need to consume. She calls this "a blatantly domestic" orientation. If it takes less time to produce what we really need (do we really need countless interchangeable Nintendos and other deliberate forms of distraction and superfluity?), then so be it. We could spend more time caring for each other, particularly as our population ages and the need for good elder and child care becomes apparent. This aspect of Yalnizyan's refreshing orientation is one that she describes as "domestic in every sense of the word."

The Reverend Jimmy Tompkins, who founded the first credit union in English Canada in 1933, was a leader of the Antigonish Movement, an organization dedicated to adult learning (as opposed to training) and co-operative economics. In the midst of the Great Depression — the crisis that gave rise to the modern welfare state — Tompkins often made the point that it isn't enough simply to have good ideas. If we are to be effective, we must have "ideas with legs."

This book is not just a critique of the dead-end course along which Canada is being steered. It is a solid effort to put new energy into the legs that will carry us in another direction.

Notes

1. Xerox advertorial published as a supplement to *The Globe and Mail*, March 4, 1993.
2. See "Canadian Paper on New Orientations for Social Policy," presented to the OECD Ministerial Meeting of the Employment, Labour and Social Affairs Subcommittee, December 8-9, 1992; New Brunswick departments of Advanced Education and Labour, Income Assistance, "NB Works: A Joint Pilot/Demonstration Project Proposal," March 1992. It is doubtful that the expensive NB Works model, which costs at least $59,000 per person for basic education and skills training, can be replicated widely given funding constraints. It is clear that some form of program that will not do similarly intensive training (but will tie benefits to a combination of work and training) is the wave of the future.
3. "Grits Vow Radical Social Reform: System Must 'Reward Effort' by Welfare Recipients, Axworthy Says," *The Globe and Mail*, Feb.1, 1994, pp. A1, A7.
4. John Kenneth Galbraith, *The Culture of Contentment* (Boston: Houghton, Mifflin, 1992), p.31.
5. Richard Barnet, "The End of Jobs," *Harper's Magazine*, September 1993.
6. Stephen Leacock, *The Unsolved Riddle of Social Justice* (New York: Jonathan Lane Co., 1920), p.145.
7. E.P. Thompson, *The Making of the English Working Class* (Harmondsworth, Middx.: Penguin, 1963), p.575.

Securing Society: Creating Canadian Social Policy

Armine Yalnizyan

Throughout the twentieth century, nations defined their responses to the heaves of the market's boom-and-bust cycles through the interplay between people's demands and their governments. The variations of that struggle were reflected in the variations among national policies; but across virtually all industrialized nations a consensus gradually emerged around the necessity for greater security to be built into the institutions of society. The architecture of post-Depression social security was built around twin pillars: the availability of stable, decently paid jobs as the primary source of income security; and the existence of a system of supports for those who found themselves outside the labour market. In hindsight the emergence of the welfare state appears as a rational response to this pattern of economic development, indeed as a precondition of its continuance.

Today many of the hallmarks of the welfare state are being dismissed as irrelevant to the era of global competition and tough economic times. The global mobility of capital, rather than the fluctuations of the economic cycle, has become the root cause of economic and social insecurity. Where once priority was placed on stability and full employment as societal goals, today the loudest call is for a "flexible" labour force with its attendant concepts of continual change in the available job/skill mix and workers' adaptability to any wage/benefit climate.

A new episode in the continuing war of conflicting values between classes is emerging, and with it new terms, and new questions. Is the welfare state an anachronism, out of step with current conditions and needs? Is it simply unaffordable? What is being offered in its place, and why? What are the real, enforceable rights of citizenship in this climate? What scope does the nation-state have to address these issues?

The Market Solution: The Twenty-Year Social Experiment

As the issues of deficit and debt have taken centre stage, goals such as full employment and the pursuit of equity among citizens have dropped off the agenda of those who shape policy, in this country and elsewhere. Regardless of political persuasion, every level of government has become ensnared by pressures to balance the books, a preoccupation that has virtually eclipsed other arenas of statesmanship. Increasingly states are embracing the market-driven approach to governing, conferring on the market the status of "neutral" arbiter of human endeavour, while setting fiscal and monetary policies that shape how the market works and rig the outcomes.

It is tempting in considering this evolution to critique the governments of the right; but the social experiment of taking the state's regulatory hands off the marketplace has been practised for more than twenty years now by governments of virtually every political persuasion. So it is relevant to ask: what has the marketplace delivered? In Canada, the answer is unambiguous:

growing inequity and instability in society at large and an increasingly precarious existence for the most vulnerable among us.

As the Canadian broadcaster Bill Roberts puts it, the culture of consumption that reigned throughout the 1980s has created a new underclass of the consumed. By 1991 4.2 million Canadians were officially poor. Of these, 1.2 million were children under the age of eighteen, representing 18.3 per cent of all Canadian children, up from around 14 per cent for most of the 1970s. While slightly more than half of all poor children live in two-parent families (54 per cent), the chances of being poor in a single-parent family headed by a woman was a staggering 62 per cent in 1991.[1]

In 1992 there were more than 2.7 million social assistance recipients in Canada, and over 1.5 million Canadians out of work. The need for food banks, perhaps the country's largest growth industry, has become a national disgrace. In 1991 2.2 million Canadians, including 850,000 children, used food banks. There are 372 communities with one or more food banks in Canada today. In 1981 there was one. Since 1990, 131 new food bank programs have opened across Canada, resulting in more than two thousand food relief outlets — more than any single restaurant or grocery supermarket chain, including McDonald's restaurants. In Metropolitan Toronto, the hub of economic activity for the nation, 45 per cent of the 162,000 (and rising) monthly food bank recipients are children, while children make up only 21 per cent of the city's total population.[2] Compared to two years ago, food bank households are now younger, more employable, more highly skilled, and better educated. What happened?

How Does Your Market Grow?
The Distribution of Income, 1973-91

Consecutive waves of recession, technological innovation, and globalization of the production methods have resulted in "creeping" unemployment. The national average rate of joblessness has grown with each passing decade:

- 2.7 per cent in the late 1940s;
- 4.2 per cent throughout the 1950s;
- 5.1 per cent in the 1960s;
- 6.7 per cent in the 1970s; and
- 9.3 per cent in the 1980s.

By 1992 the annual average rate of unemployment stood at 11.3 per cent, the highest in the industrialized world.[3]

Over the last twenty years profound restructuring has turned the labour market into an employment lottery, a huge game of musical chairs with fewer decent-paying jobs for more and more players. Episodes of declining unemployment have been characterized more by low-wage and part-time job growth than by increases in steady work at living wages, which are steadily being "shed" through periods of both recession and recovery. Faster and more widely occurring turnover between jobs (or labour market "churn"), joblessness, and the proliferation of precarious jobs are creating a growing colony of the excluded, a society of insiders and outsiders. Not surprisingly, this increased instability has resulted in more frequent resort to public forms of income assistance.

An analysis of specially requested tabulations from Statistics Canada's Survey of Consumer Finances reveals the severity of this shift. These data examine changes in pretax income of all Canadian families with never-married children under the age of eighteen between 1973 and 1991. This population has been ranked according to their average income into ten equally sized groups, or deciles. There are 358,000 families in each decile. It should be noted that, if this study had been based on the entire Canadian population (including singles and families without children) and/or "disposable" income (income minus personal, consumption, and property taxes), the reported trends would be even more pronounced.

The figures show that income disparities in the marketplace have increased steadily in Canada over the last two decades, and most rapidly in the last five years (see Table 1 and Figure 1). Between 1973 and 1987 the richest 10 per cent of Canadian families with children were the only group to significantly increase their share of market income (earnings from wages, salaries, and self-employment, and returns on investment) from 23 to 24.6 per cent, an increase of 7 per cent. Four years later, by 1991, this group had doubled their gains, with the result that the top 10 per cent now controlled 26 per cent of all market income. The average market income of families in the top 10 per cent of the population was $124,269 in 1991.

The shifting of market-generated wealth to the richest families in society is not insignificant. If the top 10 per cent's share of market income had been held at its 1973 level of 23 per cent of all earnings and returns on investment, there would be an "extra" $5.4 billion to be redistributed in 1991 alone. If that amount were equally divided among the bottom 10 cent of the population, an additional $15,124.59 would be available for each of Canada's poorest 358,100 families with children.

It has been the poorest Canadian families who have most dearly paid the cost of this transition. The 10 per cent of the population at the bottom of the heap saw a 47 per cent decline in their share of market income, to a pitiful 0.7 per cent of all earnings and returns on investment. The average market take for these families in 1991 was $3,422 a year. One explanation is that more people are being moved right out of the labour market, at least for periods of time. There is greater movement in and out of

Table 1
Distribution of Market Income among Economic Families
With Children Under 18, by Deciles, 1973 - 1991

Decile	Per Cent Share of Market Income				Change in Share (in per cent)		
	1973	1979	1987	1991	1973 - 1979	1973 - 1987	1973 - 1991
1	1.35	0.97	0.77	0.72	-28.15	-43.00	-46.54
	(1,639)	(2,239)	(3,165)	(3,422)			
2	4.18	3.90	3.27	2.56	-6.70	-21.50	-38.87
	(5,092)	(8,992)	(13,386)	(12,158)			
3	6.06	5.99	5.40	4.65	-1.15	-10.90	-23.34
	(7,380)	(13,825)	(22,100)	(22,095)			
4	7.46	7.51	7.12	6.52	0.67	-4.60	-12.54
	(9,082)	(17,326)	(29,125)	(31,022)			
5	8.65	8.77	8.59	8.27	1.39	0.60	-4.39
	(10,530)	(20,251)	(35,138)	(39,319)			
6	9.91	10.08	9.92	9.79	1.72	0.00	-1.20
	(12,071)	(23,277)	(40,559)	(46,578)			
7	11.28	11.43	11.41	11.57	1.33	1.10	2.59
	(13,737)	(26,381)	(46,658)	(55,040)			
8	12.87	13.09	13.12	13.54	1.71	1.09	5.20
	(15,678)	(30,221)	(53,644)	(64,417)			
9	15.29	15.54	15.74	16.25	1.64	3.00	6.27
	(18,624)	(35,869)	(64,385)	(77,296)			
10	22.95	22.73	24.65	26.13	-1.00	7.40	13.86
	(27,944)	(52,482)	(100,831)	(124,269)			

Notes: The numbers in brackets refer to the average family pre-tax income in each decile are in current dollars.

Market income refers to earnings from wages, salaries and self-employment plus returns on investment.

Source: Statistics Canada, Household Surveys Division, Survey of Consumer Finances, unpublished data.

the ever expanding "marginal" labour market of precarious jobs and, with more people competing for work at the edge, the barriers to full participation in society are being raised for the most vulnerable, those deemed less attractive to employers.

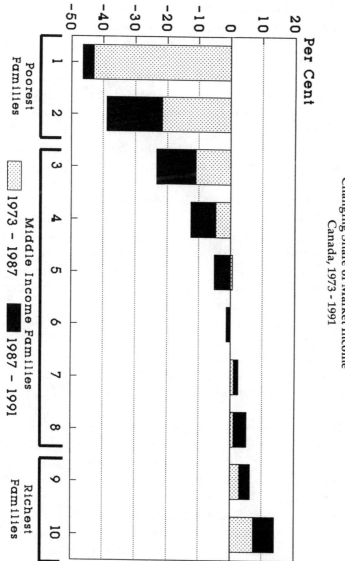

Figure 1
Changing Share of Market Income
Canada, 1973 - 1991

The only factor softening this effect is transfer payments (see Table 2 and Figure 2). Taking total income into account — which, given the focus on families with children, primarily refers to family allowances, unemployment insurance, and social assistance, but also includes pensions and old age supplements — the poorest 10 per cent of the population lost "only" 9 per cent of their total income over the nineteen-year period. The bottom 30 per cent lost, on average, "only" 11.5 per cent of their share of income. Clearly, when we compare market to total income data we can see that only income security programs have kept the poor from a free fall into destitution.

The other jarring fact of this twenty-year period is the degree to which society's reliance on the market as the primary source of income security has become a less stable proposition. For the fifteen years between 1973 and 1987, the very middle 20 per cent of families retained a relatively constant share of market wealth. Over the next five years, however, there were no groups that retained what they had had before. They had either won or lost something, and a growing proportion of the middle class was sucked into the pattern of decline. By 1991 the bottom 60 per cent of Canadian families with children had a smaller share of the market's bounty than in 1973. These changes affect families with average earnings and returns on investment of $46,578 a year or less.

But, despite overwhelming empirical evidence that markets do not function neutrally, the message is not getting through to people. Continued faith in the market "solution" is creating a deepening problem for society at large. Inequality in the distribution of income is becoming worse more rapidly, with more and more of the middle class swept into its reach. The middle class has not just been eroded. It has been destabilized, turning the country into a shifting pyramid of few winners and many losers. Only government programs have kept the lid on an increasingly

Table 2
Distribution of Total Income Among Economic Families
With Children Under 18, by Deciles, 1973 - 1991

| Decile | Per Cent Share Of Total Income | | | | Change In Share (in per cent) | | |
	1973	1979	1987	1991	1973 - 1979	1973 - 1987	1973 - 1991
1	2.33	2.04	2.19	2.10	-12.45	-5.00	-9.33
	(3,013)	(5,012)	(9,753)	(11,351)			
2	4.74	4.50	4.25	4.00	-5.26	-10.30	-15.15
	(6,124)	(11,082)	(18,987)	(21,591)			
3	6.30	6.24	5.93	5.70	-0.95	-5.90	-9.88
	(8,143)	(15,373)	(26,458)	(30,492)			
4	7.52	7.60	7.32	7.10	1.06	-2.66	-6.29
	(9,727)	(18,701)	(32,679)	(37,875)			
5	8.64	8.76	8.54	8.30	1.39	-1.16	-3.46
	(11,166)	(21,554)	(38,126)	(44,789)			
6	9.78	9.93	9.74	9.60	1.53	-0.41	-1.44
	(12,640)	(24,437)	(43,451)	(51,761)			
7	11.06	11.22	11.08	11.10	1.45	0.18	0.57
	(14,304)	(27,620)	(49,457)	(59,771)			
8	12.63	12.77	12.67	12.80	1.11	0.32	1.23
	(16,326)	(31,433)	(56,556)	(68,671)			
9	14.88	15.06	15.02	15.20	1.21	0.94	2.19
	(19,244)	(37,074)	(67,031)	(81,711)			
10	22.12	21.89	23.26	24.00	-1.04	5.15	8.70
	(28,601)	(53,890)	(103,793)	(129,176)			

Note: The numbers in brackets refer to the average family pre-tax income in each decile are in current dollars.

Total income refers to market income plus transfer payments (ui, social assistance, CPP, etc.) and retirement income.

Source: Statistics Canada, Household Surveys Division, Survey of Consumer Finances, unpublished data.

outrageous distribution of resources. It is alarming to think that these are the very programs that have been eroded since the late 1980s and that are under open attack today.

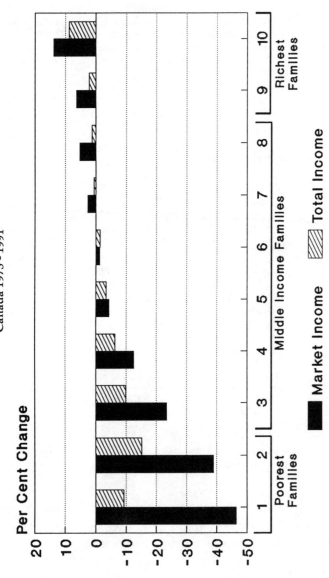

Figure 2
Changing Share of Income
Canada 1973 - 1991

Shifting Time: The Distribution of Working Hours

Money is not the only thing that the marketplace has redistributed over the last twenty years. Since 1970 Canadians have been producing more with less, which is a testament to labour-saving technologies certainly, but also to the way working time has been reorganized.

Between 1970 and 1990 Canadians more than doubled the nation's wealth (taking inflation into account) by using only 50 per cent more labour (see Table 3). The economy grew most rapidly in the early 1970s, but that leap of 29 per cent required almost 14 per cent more working time. By the early 1980s an economy surging ahead by more than 15 per cent only required 4 per cent more labour. Throughout the period the growing value of production far outstripped how much working time was needed to generate this wealth.

This trend was most pronounced in the manufacturing sector, where between 1975 and 1990 the real value of production grew by 26 per cent, with only a 5 per cent increase in time spent actually processing materials. (The industrial classification "manufacturing" also includes people working in white-collar occupations such as sales and administrative functions. "Processing" is an occupational classification and more accurately captures the amount of time required to get the product made.) Unquestionably, there has been a sizeable productivity dividend; but since the mid-1970s workers have not tended to get much of a share of it. This can be seen by the relative stagnation of the purchasing power of median family incomes since the mid-1970s, as shown in Figure 3. The treadmill feeling of dual-earning families is no illusion: even though more members of the family are working, most families have been unable to significantly improve their lot. In fact, they are feeling more insecure.

Just as the distribution of wealth is becoming more unequal across society, so too is the allocation of working time becoming

Table 3
Time and Money:
The Real Value of the Canadian Economy, 1970 - 1990

Value of Production (in 000s, constant $1986)

	1970	1975	1980	1985	1990	1991
All Economy(gdp)	271,372	350,113	424,537	489,437	567,541	558,862
Manufacturing	149,814	197,971	247,318	252,578	249,591	219,481
Mfg as % of Gdp	55.2	56.5	58.3	51.6	44.0	39.3
Change (in %)	**1970-75**	**1975-80**	**1980-85**	**1985-90**	**1990-91**	**1970-90**	**1975-90**
All Economy	29.0	21.3	15.3	16.0	-1.5	209.1	62.1
Manufacturing	32.1	24.9	2.1	-1.2	-6.5	66.6	26.1

Actual Hours Worked (in 000s of hours)

	1970	1975	1980	1985	1990	1991
All Economy	291,015	330,826	371,658	386,363	438,589	429,603
Manufacturing	NA	67,649	76,096	74,991	71,932	67,982
Processing	NA	53,800	60,098	56,077	56,437	53,132
Mfg. as % of Gdp	20.4	20.5	19.4	16.4	15.8	
Change (in %)	**1970-75**	**1975-80**	**1980-85**	**1985-90**	**1990-91**	**1970-90**	**1975-90**
All Economy	13.7	12.3	4.0	13.5	-2.0	50.7	32.6
Manufacturing	NA	12.5	-1.5	-4.1	-5.5	NA	6.3
Processing	NA	11.7	-6.7	0.6	-5.9	NA	4.9

Sources: Gross Domestic Product (gdp) figures in constant (1986) dollars, from Table H3, Bank of Canada Review; manufacturing figures from Statistics Canada Catalogue 31-001, Inventories, Shipments and Orders in Manufacturing Industries, deflated by the Consumer Price Index; Statistics Canada 71-001, The Labour Force.

increasingly skewed. Table 4 shows a clear trend away from full-time work as the standard to an increasing proportion of people working either part-time or very long hours. While the format for recording working hours was different in 1971, it appears that earlier data reinforce this trend: out of a total 8.15 million people actually working some hours (that is, excluding people on temporary layoff or holiday), 8.3 per cent worked over fifty-four hours and 16.3 per cent worked twenty-four hours or less a week. In between, 75.4 per cent of the working population logged between twenty-five and fifty-four hours a week.

Figure 3
Median Family Income ($1989)
Canada 1965-1990

Thousands

—＊— Median Family Income

Source: Income Distribution by Size

Since 1985 the most remarkable trend has been the increased use of part-time workers regardless of an overall increase or decrease in total hours worked in the economy. Quite apart from growing joblessness, the phenomenon of *under*employment, at this most basic quantitative level, reinforces the trend to falling income for many people.

Table 4
The Distribution of Actual* Work Time, Canada, 1976 - 1992

| | Number of workers (000s) | | | |
	1976	1985	1990	1992
Part-time (1-29 hours)	1,483	2,073	2,378	2,431
(% of total)	16.9	19.8	20.4	21.7
Full-time (30-49 hours)	6,151	6,868	7,399	7,110
(% of total)	69.9	65.6	63.5	63.4
Full-time (50 or more)	1,164	1,522	1,868	1,699
(% of total)	13.2	14.5	16.0	15.2
Total	8,798	10,463	11,645	11,210
Total persons**	9,477	11,311	12,248	12,240

* Refers to total of all hours worked at all jobs in reference week, i.e. actual hours worked.

** Total persons includes people working 0 hours in reference week.

Source: Statistics Canada, Labour Force Annual Averages 1975-1983; The Labour Force, December 1984, December 1985, December 1990, December 1992.

The Evolution of Canadian Social Security

The current crisis in social security is made more understandable by tracing its history. The evolution of the way people and their governments have constructed the intersection between market and non-market experience has already had three discrete phases in Canada, and it appears we are about to embark on a fourth.

First, there is the prehistory of Canadian social security, before the Depression, a time when unpaid labour played a much greater role not only in families but also in whole communities. In this sense, Canadian society was less linked to the market. But it was still a "tough-luck" scenario for those who did not have gainful employment, or someone to support them. The only mechanisms for help were local and voluntary sources of relief (and "make-work" projects) such as mutual aid circles, religious and civic associations, extended families, immigrant ghettos, and neighbours; and, in urban centres, "indoor relief" in the forms of workhouses and prisons.

As minimal as these forms of support were, they were more available then than today because people were less mobile and more likely to live in rural communities or tightly defined urban neighbourhoods. This made the links between people more obvious and enduring. Capital, too, was familiar and local, largely created and used within narrow geographic confines. The significance of this manifestation of capital will be revisited throughout the chapter.

Forging a New Language of Social Security

The visceral experience of the Great Depression, followed by an unparalleled experience of creating an economy together to fight the war, created a space for acceptance of a new concept: public welfare, in the sense of the general well-being of people within a society. If the Depression was widely felt, it was still class-based. The war effort, on the other hand, was a truly collective exercise, cutting across all classes, indeed across nation-states. It became abundantly clear that economic outcomes were not God-given, that the seemingly inexorable logic of the marketplace was a myth. The shiver of universal risk had swept over everyone, and people started demanding protections by pooling that risk across society, and not just at the traditional levels of municipalities and provinces. For the first time, people sought a larger scale of protection, at the highest jurisdictional level, to provide certain minimums.

This experience joined forcefully with the nervousness of politicians and citizens alike, who confronted the seeming inevitability of a new recessionary wave stemming from the lags in the postwar reconversion of the economy. That combination — hardship fresh in people's minds, the knowledge that the state had potent levers at its disposal, and an apparently imminent economic nosedive—predisposed the Canadian government, and governments elsewhere, to intervene in peacetime in a wholly new way.[4]

This was not simply a case of enlightened leadership. The impact of Russia's 1917 revolution had rippled through the ruling classes of the industrialized nations for decades, a constant reminder of what could happen at home. In Canada during the 1930s the possibility of widespread anarchy and revolt in the streets was never more real. It was a time of imaginative fight-back tactics, plant seizures, and the historic "On to Ottawa" trek of 1935, when thousands of relief-camp workers from British Columbia and the Prairie provinces rode the rails towards Ottawa. They went to protest joblessness and the wages and conditions of the work camps, which "recruited" single employable males through vagrancy laws to clear forests and build roads in remote areas for twenty-five cents a day. They went to demand closing of the camps and creation of both public works projects paid at union scale and unemployment insurance for the unemployed. They were forcibly dispersed in Regina on July 1st by the RCMP, who opened fire on them, leaving a rump of unemployed protesters to make their way from Toronto to Parliament Hill on foot.

Such actions had their impact. As early as 1937 the federal government was emphasizing the need for "collective efforts to promote economic development and collective assumption of the responsibility for the alleviation of individual distress." The Rowell-Sirois Commission stressed that "rising standards of public welfare and education had come to play an immensely important part in the economic affairs of the country."[5] This perspective eventually consolidated the rights of respective governments to their "fair" share of national resources in order to discharge their responsibilities to "a) promot[e] equal opportunities for the well-being of all Canadians; b) [further] economic development to reduce disparity in opportunity; c) provid[e] essential public services of reasonable quality to all Canadians."[6]

In the context of growth and a relatively broad social consensus, from the end of the 1930s to the beginning of the 1970s Canadians pushed and shoved a new definition of themselves out of national institutions. With it came a new, evocative language of citizenship, a language of rights to the provision of certain protections and of access to services. The expansion of individual rights was couched in two powerful concepts: one of uniform protection across the country, essentially giving concrete form to national citizenship; and the growing notion of universality of certain minimums, albeit basic and categorical, "obtainable as of right and in the company of all other citizens."[7]

The concepts of collective and state responsibility to improve things had taken hold and laid the groundwork for what Canadian citizens could expect from their governments, no matter where they lived. All the major milestones of the Canadian welfare state — old age security (1927), unemployment insurance (1940), the family allowance (1945), the formalization of equalization payments (1957), the Canada Pension Plan (1964), the Canada Assistance Plan (1966), and medicare (1966) — reinforced the shift in responsibility for defining the scope of the welfare state from provincial governments to the national government.[8] The backdrop for these developments was the explicit assumption that people would be working and there would be jobs for all who wanted to work. As the 1943 Marsh report put it, "The starting point of all social security discussion is the level of family income or . . . wages."[9] Social security without stable employment was simply not possible.

Clearly, policies for economic and welfare security buttress each other. But what props the twin pillars? The foundation of this social architecture is the availability of capital, both to invest in productive activities that create a stock of jobs, and to provide a revenue base, through taxes, to finance social programs. During this period capital was less localized than before but still largely created and used within the confines of the nation, and this is key to the type of policies that could be implemented. The Marsh report articulates the interrelation of all three elements of the system (capital, jobs, income support) and the precondition of its smooth functioning — the formulation of a low interest-rate policy.[10]

Rethinking the Meaning of Social Security

At least two currents of thought began to interrupt these developments. First, there was an emergent sense that the bedrock for these structures — high levels of employment at living wages throughout the nation — was not automatically part of a growing economy in macroeconomic terms. High unemployment levels (around 7 per cent from 1958 to 1962, compared to about 3 per cent from 1946 to 1956) and the 1950s wave of technological change had raised the issue of structural mismatches due to skill shortages in the labour market, a situation that could create the anomaly of high unemployment and jobs going begging.[11] The Keynesian assumption that government stabilization policy could soak up this unemployment was placed in question, and by 1957 the federal government had quickly taken up the themes of training and upgrading skills. Direct job creation was almost a throwaway idea, in the form of a marginal program of "Winter Works" introduced in 1958 offering a 50 per cent wage subsidy to employers for municipal construction undertaken in the winter months.[12]

Second, the atmosphere of the 1960s was infused with "notions of post-industrialism, the social service state, the need to encourage meaningful work experience and socially productive jobs, community development and participatory decision-making structures."[13] There was a sense of impatience with the gains of the day, and in particular a sense that the state's response to the problems of poverty in Canada was wholly inadequate. This was the era of Lyndon Johnson's "war on poverty," Trudeau's "Just Society" campaign, and the proliferation of welfare rights groups continually organizing, demonstrating, and developing agendas for parliamentary and extra-parliamentary action. In 1971 hundreds of groups from across Canada came together at the First Poor People's Conference to draw attention to the situation of the poor, share agendas, and develop new tactics.

This serves as a reminder about where the social momentum for change comes from. In the context of a broad-based acceptance that there was too much inequality in society, the mobilization of the disenfranchised generated ideas that shaped the policy debate. By 1966 the Canada Assistance Plan had consolidated the public assistance system, but it did not formulate enforceable standards of assistance across the country. Just two years later, the Senate Committee on Poverty was mandated to investigate all aspects of poverty and recommend effective policy measures. When its findings were presented in 1971 (the Croll report), it recommended that these programs be scrapped and replaced by a guaranteed annual income scheme.[14]

The preoccupation with finding the right response to poverty was renewed in the Federal-Provincial Review of Social Security, initiated in 1973 with the release of the *Working Paper on Social Security in Canada* (commonly known as the Orange Paper), produced by the Minister of National Health and Welfare, Marc Lalonde. The Orange Paper was both the final flourish of the first era of social security reform in Canada, and a harbinger of the next phase. It stressed the need to meet community and individual needs through work that was socially useful, and it defined policies that could combat poverty to produce a fair distribution of income between people. The Orange Paper formulated a two-tiered approach to social assistance (while preserving and expanding forms of social insurance)—a guaranteed annual income scheme for those who could not work, and an income supplement for the working poor.

The Social Security Review, which lasted from 1973 to 1976, was the last gasp of a broader view of income security reform. It was widely seen as a failure. The most notable product derived from the process was the 1978 introduction of the Refundable Child Tax Credit, basically a limited income supplementation scheme created with funds already in the social welfare system. It was a sad summary of how dramatically the tide had turned. What would have been dismissed as tinkering a few years earlier was heralded as one of the most important changes in Canadian social policy. It represented the triumph of incrementalism. Noodling with the tax system and revenue-neutral programs of income supplementation had replaced any notion of guaranteeing a social minimum for citizens. The era of expanding "universal" rights was fading fast. In the wake of broad-based activism came a new attitude and approach through campaigns devoted to narrower interests: the women's movement, senior citizens' groups, the Native community, environmental groups, and coalitions of visible minorities struggling for employment equity, pay equity, old age security improvements, child care, long-term care, parental rights, self-government, and a cleaner planet.

But the mid-1970s also marked a significant shift in the state—and public—orientation to social security. By 1975, for the first time in the postwar period, a federal deficit emerged that would not go away.[15] It was portrayed as the product of runaway social expenditures, but the genesis of the deficit was elsewhere: in deliberate government policies, a simple case of revenues not keeping up with expenditures... by choice (see

Figure 4). What has been referred to as the "fifth wave" of debt creation in Canada began as a result of changes in tax policy, changes that led to a decline in total tax revenues relative to the size of the economy, thus raising the requirement for borrowing.[16]

Figure 4
Federal Revenue, Expenditure, Deficit
As Per Cent of GDP, Canada 1961-1991

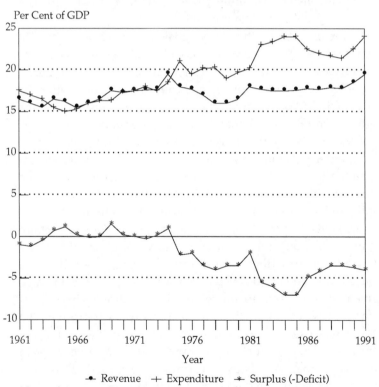

Per Cent of GDP

Year

— Revenue   + Expenditure   ✳ Surplus (-Deficit)
National Accounts Basis

Relative to GDP growth, expenditures remained flat throughout the 1970s, responding only to the crisis of 1981-82. The mood of fiscal restraint that had emerged by the mid-1970s was parallelled by a mood of social conservatism, partly a backlash to the controversy sparked by the massive broadening of UI entitlements in 1971. A new discourse flowered, one that branded expenditures on social security as disincentives to work. In the public mind welfare was losing its association with the well-being of citizens and beginning to resonate more frequently as a drain on the public purse and individual enterprise.

Discussion about the lofty goal of guaranteed annual incomes (at least the progressive model of raising the social minimum in order to combat poverty) withered, but the language and policies of work incentives had caught fire. Shortly after the launch of the Orange Paper the provinces established various programs, largely in anticipation of a broader federal restructuring of social security that never materialized. With policy initiatives splintering along regional lines, the federal government suspended its role as standard-setter for the nation and became more of a "blind banker." In 1974 Manitoba introduced MINCOME, an income-maintenance experiment for employable welfare recipients. In 1978 Alberta defined single mothers on welfare as employable, and therefore subject to work requirements. By 1979 both Ontario (the Work Incentive Program—WIN) and Quebec (Supplément au revenu de travail—SUPRET) had wage supplement programs to move social assistance recipients off welfare. WIN was time-limited to a two-year transition; the SUPRET was used as the symbol of Quebec's goal to establish a guaranteed income; but both programs were modest ventures in public spending, and both governments expected to recoup funds due to savings on the assistance bill.[17]

The various models of income supplementation represented a reduction in guaranteed levels of assistance for employable welfare recipients.[18] After all, this was the heart of providing incentives to work. The stage was set for a perverse reorientation of policy, in which those on assistance, in particular welfare mothers, were to be reformed instead of the government reforming the terms of access to work and work itself. For by now an emerging marginal economy was noticeable. The cycling of people between social assistance and work can only be understood in the light of this segment of labour market opportunities, characterized by low wages, poor working conditions, irregular hours of work, and, therefore, high turnover.[19]

Rewriting the Codes of Social Security

One could look at the years 1976 to 1985 as the period of a draw. Competing visions of the appropriate policy route often ended in gridlock, precisely because "to encourage work, to reduce poverty and to contain costs are incompatible aims."[20] This incompatibility was soon swept away by the supremacy of containing costs. Between 1981 and 1982, with the first jolt of the recession, the federal deficit jumped from $7.3 billion to $20.3 billion. It peaked at $31.4 billion in 1984, then started to fall back, but never again dropped below the $20 billion mark.[21] The election of the Conservative government in 1984 transformed this dilemma into mythic proportions, and public opinion passed through the looking-glass. Within a decade the war on poverty had been supplanted by the war on the deficit. It provided the perfect cover for battling expectations for economic justice among workers and the poor.

Again, these changes did not originate within government, but rather were orchestrated by governments in response to extraparliamentary forces. In 1976, 150 of the country's largest, mostly multinational, companies came together to form the Business Council on National Issues. Its CEOs were devoted to addressing issues of national interest, not specifically economic ones. By the early 1980s, in part due to the recession, there was no semblance of broad social consensus on issues of social security. Filling this vacuum, the spokesmen of the business sector grew increasingly articulate and aggressive with their program for change. In a period marked by the proliferation of royal commissions and parliamentary sub-committees, the corporate executives were given ample venues for influencing public policy. The big finger of business was most evident in the work of the Macdonald Commission on Canada's prospects for the future, launched by the Trudeau government in 1982 and reporting to the Conservatives in 1985. Predictably, the agenda for social security reforms began to take on a radically new shape.

Policies that stayed the course on the accumulating debt, high inflation and high interest rates, and tax trends were treated as "exogenous variables," the unmovable parameters within which public discussion about social policies would be permitted. In this climate it became possible to launch an ideological attack on the government's statutory obligations to help those affected by the fall-out from the 1981-82 recession. Between 1982 and 1991 the public-debt charge was the single fastest growing category of federal expenditure in both absolute and relative terms, increasing by 79 per cent in real terms over the period. (See Table 5. Note these figures are in 1986 constant dollars, not current dollars.)

The second most rapid growth was for the second largest expenditure item, the national program of unemployment insurance (71 per cent over the period). This was an unexpected finding given that the federal government had reduced both eligibility qualifications and benefit periods for the unemployed in 1990. This increase merely underscores the severity of the current crisis in the labour market. The third most rapid growth was for CAP, which finances social assistance and is a comparatively

Table 5: Expenditures, Revenue and GDP (in 000,000s, $1986), Canada, 1982 - 1991

	1982	1983	1984	1985	1986	1987	1988	1989	1990	1991	% Change 1982-1991
OAS, GIS, Spouse Allowance	11332	11475	11759	12331	13026	13445	13775	13986	14216	14390	26.98%
UI	7189	11689	11054	10856	10437	10444	10068	10094	10291	12319	71.36%
Family Allowance	2666	2655	2628	2611	2601	2534	2461	2398	2335	2298	-13.81%
Other	1666	1511	1330	1429	1586	1393	1450	1468	1358	1329	-20.23%
Total Transfer Payments											
To Individuals	22853	27331	26771	27228	27650	27816	27754	27946	28199	30336	32.74%
Ins. and Med Care	5654	4831	6287	6836	6656	6607	6296	6144	5863	5068	-10.36%
CAP	3033	3370	3715	4045	4073	4051	4076	4192	4405	4862	60.28%
Education support	2149	1823	2333	2446	2368	2232	2152	2049	1906	1563	-27.26%
Fiscal arrangements	6270	6660	6754	6464	6179	6302	6727	7477	7615	6955	10.93%
Other	1304	1288	1388	1672	1794	1461	1871	2295	2326	1966	50.78%
Total Transfer to Other											
Levels of Government	18410	17973	20478	21463	21069	20653	21122	22156	22115	20415	10.89%
Public Debt Charges	19950	20115	20427	24251	26459	26658	27867	30515	34162	35731	79.10%
Total Net Expenditures	100175	106732	109706	118333	115977	116664	120780	122377	125579	125976	25.76%
Share of Revenue, By Source											
Personal Income Tax	43.5%	47.2%	47.4%	46.1%	48.4%	49.6%	51.8%	49.6%	50.4%	54.0%	24.14%
Corporate Income Tax	14.6%	12.8%	12.8%	14.8%	13.5%	12.9%	12.5%	12.6%	12.6%	11.0%	-24.66%
Sales & Other Taxes	30.6%	30.3%	30.3%	30.4%	30.3%	29.5%	27.9%	29.8%	28.9%	26.1%	-14.71%
Non-Tax Revenue	11.5%	9.7%	9.5%	8.8%	7.8%	7.9%	7.9%	8.0%	8.1%	8.9%	-22.61%
Total Revenue	79605	72188	72510	76740	80010	85931	93708	95742	100062	100257	25.94%
Gross National Product	425970	439448	467167	489437	505666	526730	551423	564990	567541	558862	31.20%

Sources: Public Accounts of Canada, 1990-91, Vol. 1, Table 1.7; Bank of Canada Review, June 1992, Table H3.

small expenditure outlay for the government. Although the Canadian economy was almost one-third larger after the boom of the 1980s, social spending grew by only 24 per cent between 1982 and 1991.[22] By 1991 the proportion of GDP devoted to these expenditures had fallen to 7.5 per cent. Clearly, the story of Canadian debt is not about overblown social spending, but about a political unwillingness to recapture lost revenue. The most significant cause is the systematic reduction of effective corporate income tax rates, which has a long history in Canada (see Figure 5).[23]

But the chorus chanting that people, not economics, had to reform grew louder and more persistent, resulting in statements such as, "Social policy must facilitate and assist the occupational, industrial, and even geographical relocation that the new world economic order is requiring of the present generation of Canadians."[24] Between 1985 and 1989, major developments in labour market policy focused on supply-side measures: training and education, mobility assistance, self-employment initiatives, work placements, and increasing incentives to work for social assistance recipients through breaks in the tax structure. As late as 1988, Prime Minister Mulroney was citing the Ontario government's Social Assistance Review Committee's report, saying, "The best income security is a job." But his battle cry of "jobs, jobs, jobs" was reduced to "training, training, training." There was, and continues to be, stony silence about the demand side of the labour market equation: where are the jobs for which people are training?

Other policies echoed the stinging irony of getting the poor to pay off the costs of a debt created disproportionately by corporations and the rich. Tax reform took on new meaning, with an agenda based on broadening the revenue base at the bottom and reducing rates at the top. Between 1987 and 1989, the ten brackets of personal income tax rates were collapsed into three.

Figure 5
Federal Revenue by Source*
1951-1991

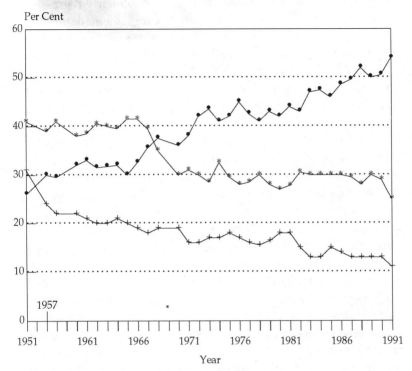

── Personal Income Tax ─+─ Corporate Income Tax ─*─ Other Taxes
* Excludes Non-Tax Revenue which accounted for 8.9% of Federal
   Revenue in 1991
Source: Public Accounts

De-indexation of the system, starting in 1985 but building momentum through various additional measures, consistently hurt the poor the most. Between 1984 and 1991 the various changes made to the personal income tax system resulted in a 386 per cent increase in the income tax burden for a low-income couple with two children (earnings of $20,000); a 15 per cent increase for a middle-income family ($50,000); and a 4 per cent increase for an affluent family ($100,000). A similarly regressive change came into effect in 1991 with the implementation of the Goods and Services Tax (GST), hailed by the federal government as the chief new mechanism of debt control. This consumption tax subjects, by design, only the lowest-income Canadians to automatic sales tax increases over time as sales tax credits that have been available to the poor shrink in value through their de-indexation.[25]

The years 1985 to 1990 have been dubbed the era of "social policy by stealth." In a remarkably short period, several key components of the social safety net were significantly changed or eroded. The 1985 partial de-indexation of family allowances and child tax credits meant that $3.5 billion less reached the pockets of Canadian families between 1986 and 1991.[26] In 1989 the government introduced a "clawback" of child benefits, applying a special taxback on family allowance payments for families with net incomes over $50,000 (not indexed for inflation). This dealt a body-blow to the notion of universality, one of the supposedly sacred tenets of the Canadian welfare state. It also laid bare the government's preference to tax one form of income differently than another. For example, because of the clawback an affluent family can have an implicit tax of 100 per cent on child benefits of $400, but their $4,000 worth of capital gains is not taxed at all.

In 1989 the federal government began to withdraw openly from its role of mitigating the effects of an increasingly unstable economy. That year the government unilaterally restricted its share of funding to provincial governments for cost-shared social assistance programs, the majority of whose beneficiaries are families with children, an act which was initially presumed illegal given the governing legislation. The decision was appealed to Canada's Supreme Court but eventually upheld. By

1992 the three provinces affected by this decision (Alberta, British Columbia, and Ontario) had been denied federal revenues to the tune of $4 billion.

In 1990 the national unemployment insurance system was gutted when the federal government withdrew its funding of the program, restricted eligibility for benefits, reduced duration of benefit entitlements, increased penalties on those disqualified or disentitled, raised premiums, and converted up to 15 per cent of the fund for "developmental uses," which include (but are not exclusively for) training purposes. At the same time the government steadily reduced allocations for training from the general tax revenue base, the Consolidated Revenue Fund, so that in real terms (constant dollars) these expenditures decreased by almost half between 1985 and 1992.[27]

As a result of these moves, the unemployed who are no longer eligible or who have exhausted their benefits now swell the ranks of a social assistance system that is increasingly underfinanced. The consequent mounting "fiscal stress" on the system has exerted pressures to further cut resources (through government targeting) and to find ways of reframing policies to reduce government responsibility for those outside the labour market. The themes of privatization and devolution, reinforced by the constitutional agenda, are reversing the developments in social security that took place in the 1930s. We are going backwards, mirroring a time past. Costs, and initiatives, are being downloaded from the federal to provincial to municipal levels of government. The same process is also shifting responsibilities, and expectations, from the public to the private spheres. This is happening in several ways: from public sector to private sector, both in terms of business and the voluntary sector; and from the public purse into the pockets — and on the backs — of those least able to bear these costs.

In a few short years the federal government successfully cut entitlements right under people's noses by conducting a highly technical and complex campaign of reform. At the same time Ottawa was doing new takes on Canadian shibboleths, making people question the very relevance of those entitlements. The most successful of these attacks was directed at the necessity of universality ("Does a bank president's wife really need a monthly family allowance cheque for $33?"); while inserting the notion that selectivity and targeting the "deserving" poorest form a more humane and rational response to the dilemmas of Canadian society.[28]

The 1985 report of the Royal Commission on the Economic Union and Development Prospects for Canada (the Macdonald Commission) was seminal in this process. It took the idea of a guaranteed annual income, last seen during the days of the Social Security Review in the early 1970s, and ran with it.[29] The report recommended radical surgery of the social security system, which was to be replaced by a Universal Income Security Program. It presented two options, both of them at remarkably low levels. But the explicitly favoured design was a guaranteed annual income of $2,750 per adult and $750 per child. To finance this the government would have to eliminate family allowances, child tax credits, married exemptions, child exemptions, the guaranteed income supplement for the elderly, federal social housing programs, and federal contributions to the Canada Assistance Plan. The Commission also called for "less emphasis" on minimum wage laws and for "trimming" the unemployment insurance system. A decade later the agenda has a familiar ring. With the exception of income supplements for the elderly (a political no-sell) and CAP (hold your breath), this is the shopping list that has been systematically ticked off in the 1990s. The stage was set for a new struggle to define social security in Canada.

Race to the Bottom: Agenda for the Future

The day after the constitutional referendum in the fall of 1992, the Canadian federal government announced its economic remedy for the troubled nation. Although not immediately obvious, the connection between the two major planks of this program of action—implementing legislation for the North American Free Trade Agreement and a proposal to overhaul income security programs—is profound. Income security has become handmaiden to the era of global competition—but not just any form of income security. The approach taken must reinforce one's primary link to the world through providing one's labour services.

This is to be achieved by handcuffing people ever more tightly to the labour market, which increasingly demands a twenty-four-hour-a-day, seven-days-a-week, on-call-as-needed commitment. The propelling force behind getting people to accept this model is a pervasive sense of insecurity and rapid change. In the labour market unreliable hours of work and income become more commonplace, and competition for existing work intensifies both locally and with workers far away.

The heightened level of competition in money markets and the creation of "rootless" capital underscore the hostage-taking of national labour markets in two ways: through the continuing separation of speculative and productive investment capital (a short-term versus long-term split in the use of money), with growing scarcity of productive investment capital; and through the virtually instantaneous movement of huge volumes of capital around the world due to advances in telecommunications and the development of new financial "tools." The national government's role reinforces these trends by eroding the notion that people might be able to step outside of the labour market, even if only briefly, with assurance of some source of income.

Name this agenda "There are no guarantees in life." Its central message is "Your job is your only salvation."

Defining the Deserving Poor

The dual-labour market, a term first coined in the 1970s to depict a split between peripheral/core (or marginal/secure) sectors of job opportunity, has become reinforced throughout the 1980s and into the 1990s. The "marginal" labour market is expanding rapidly, creating a proliferation of precarious jobs that mirror the just-in-time system of production with a just-in-time labour force.[30]

Although this model of the labour market originated in manufacturing industries, it now echoes throughout the entire economy, in retail sales, data and word processing, health care, and professional services. It is a response to businesses characterized by peak periods of production, tight turnaround times on contracts, and sporadic demand. At the same time an insider-outsider model of interaction with the labour market has developed as economic recovery becomes more associated with jobless growth.[31]

These trends create growing pressure to make up for the inadequacies of the market. Authorities portray the current system of social security as an expensive albatross, precisely because its design was based on the assumptions that: 1) the vast majority of citizens could rely on the job market as their primary source of income security; and 2) for those who were not supported by this system, the public purse could capture significant revenues from capital to finance the safety net. Both of these assumptions are evaporating in today's economic dynamic.

The government's proposed overhaul of income security should contain few surprises. "Streamlining," "fairness," and "simplicity" have become code words for further targeting and cutbacks. Tired rhetoric about work incentives is the centrepiece of reforms that refuse to acknowledge that people can't scrounge up a job. The categories of deserving are photogenically ranked — working poor, with children, and so on. Cascading tiers of support will depend on how "attached" one is to the labour market. We have been confronted with the essential thrust of these changes for some time now. But three recent events at the federal level show how far from complete this process is: introduction of the Child Tax Benefit in 1993; the huge (if as yet undeserved) hoopla over the self-sufficiency pilot projects in British Columbia and New Brunswick; and health and welfare minister Benoît Bouchard's speech to the OECD in December 1992.

Just in time for the election year of 1993 the Child Tax Benefit was offered as the prototype of an income security model based on the tax system and designed to promote the "work ethic." The heart of the new package, and central to the government's vision of income security, is the earned income supplement. The program avoids the issue of growing joblessness and poverty stemming from unemployment. It offers a whiff more income for those who already earn some, but nothing to the poorest in society, the growing numbers of those outside the labour market. Its target is the working poor with dependent children. Households that earn between $3,750 and $20,921 a year are given a wage supplement that peaks at a handsome $500 a year.

The similarity to the earned tax credit in the United States is striking. The U.S. program provides a refundable tax credit to the "deserving" poor, working poor with children. In 1990 about 10 million families earning less than $20,264 received an average credit of $570. Although meagre in its support levels, the program is consistently used to forestall arguments for improvements in U.S. minimum wage and unemployment insurance entitlements. The discussion in Canada promised to ring the same bells, jangling reminders of the Macdonald Commission's recommendations.

The Self-Sufficiency Project — Refining Income Supplementation

The second window to the future was opened late in 1992, when the federal government launched two pilot projects, one in British Columbia and one in New Brunswick, designed to "enhance the employability" of social assistance recipients. Although evaluations of the project's effectiveness — either from the perspective of saving welfare costs or moving people out of poverty — will not be submitted until the year 2000, both the federal Health Minister of the day and some provincial premiers hailed it as the shape of things to come. While each project offered a training component as a potential option, what the politicians were talking about was the "Self-Sufficiency Project" (SSP). The new federal Liberal government appears just as taken by the approach.

SSP offers no new ideas about welfare reform, either in principles or targets. It simply takes the design of income supplementation to a further extreme. The supplement is offered for a limited three-year period, to a target population of single mothers who have been on social assistance for more than twelve months.[32] There is no systematic process for potential recipients to find out about the program (it could happen in a welfare office, it could be through direct or indirect referral, it could happen by phone or a personal home visit), but once informed about it individuals are given the "option" of filling out a form certifying their eligibility for the supplement. This fulfils the "voluntary" criterion of participation.[33] Once these women have been notified that they can enrol in the program, they have up to twelve months to find themselves a job that fits the terms of the program, namely permanent and full-time (defined as at least thirty hours a week). Participation in the program means switching off welfare once the job starts, which in turn means leaving behind all supports that the social assistance system might offer, such as bus passes, clothing assistance, or coverage of drug, dental, or child-care costs.

The supplement, paid on a monthly basis, is calculated at 50 per cent of the difference between actual annual earnings and an assigned ceiling of annual income.[34] It is designed to be largest for a woman working thirty hours at the minimum wage and drops thereafter; higher total income comes from working longer hours and/or getting a job that pays higher than minimum wage. In New Brunswick the supplement peaks at $8,600 for a woman earning the minimum wage of $5, which for a full year at thirty hours a week would gross her $7,800. Combined with the maximum supplement this single parent would, for a limited period, have a total income of $16,400 to cover all her and her children's expenses. The supplement stops if the woman fails to meet the work requirement during two or three successive months (that is, if she drops below an average of 120 hours per month, regardless of reduction in hours, sick time, and vacation time). Of course, her market income would have also dropped. The size of the supplement varies only with income earned and is consequently of most value to smaller families.

Income supplementation has become the central weapon in the various governments' arsenals aimed at lightening the welfare load. It represents one translation of the shift from "passive" to "active" government interventions. Demonstrated need is no longer enough; providing income support for those who meet the eligibility requirements is laudable, but the government suggests it is in people's own interest to impose additional requirements (for training and working) in return for help.

Income supplementation has also become the key policy

concept for dealing with the margins of the labour market—jobs that do not pay living wages. This is the mechanism, rather than a guaranteed income, that will provide the rationale for under-cutting the validity of the minimum wage. But supplementation cannot provide a similar type of earnings "floor" for two reasons. First, the programs are always small scale and precisely targeted, so that in no sense can they provide a guaranteed floor for the working poor. They are only for some well-defined groups of people and for limited periods of time. (Remember, these pilot projects are designed as "loss leaders"—most generous at the outset to incite support.) Second, one of the features of the SSP design is that the supplement is geared to a certain target, a lower maximum supplement being tied to a lower target income; so in this version, supplementation decreases as lower wages become the norm.

If labour markets continue to develop along the lines of the last decade, income supplementation will be increasingly looked to as an important "transitional" measure, even though it simply reinforces the problem. As programs of income supplementation are taken up more widely, an increasing proportion of the public purse goes to support a growing margin of the labour market that does not receive a living wage. Meanwhile, the public purse is less able to capture revenue from hypermobile capital, and the population is generating less revenue through a combination of lower median incomes and more frequent and/or longer bouts of joblessness. So supplementation itself eventually becomes perceived as too costly and is gradually reduced through the "salami principle" of cutting back one thin slice at a time, or through increasingly refined targeting and tightening of eligibility criteria. Eventually it will be abandoned or reduced to an aftertaste of a policy, irrelevant to most people despite their need.

The New War of Words

The third key to understanding the road ahead is in the text of
Canada's message to the Organization for Economic Co-
operation and Development in December 1992.[35] The OECD
speech is filled with references to "promoting self-reliance,"
"breaking the spiral of dependency," and "enabling self-
sufficiency" in order to "more fully participate in the workforce." In the section on future directions, it trots out the old
chestnut about "targeting scarce resources." Then it moves on to
the punchline:

> There must be a re-examination of values to focus more on
> achieving *equality of opportunity*, not on inequality of incomes
> alone. It is important that the reorientation of social policy be
> *forward looking* and not be caught in a debate on yesterday's
> solutions to yesterday's problems. For example, the long-
> standing proposal for a comprehensive guaranteed annual
> income (GAI) for all members of society fails to recognize that
> income support is not the solution to the root problems faced
> by many of those who find themselves in need of support.[36]

It is disingenuous, maybe even meaningless, to suggest that
equality of outcome is less important than equality of opportunity. The outcomes of one generation (such as growing inequali-
ties in income) are integrally linked to the opportunities of the
next. What happens to and within families affects how their children face the future.

But at this moment in history everything, including revision
of the welfare state, serves at the altar of deficit reduction. By
today's standards, even the niggardly guaranteed income proposed by the Macdonald Commission seems wildly utopian.
Continued cutting back is the only option, according to politicians of every stripe.

The death of real thinking about anything but the deficit has
led to the death of policy. What is offered in its place is a

renewed emphasis on employment, increasingly absurd as access to employment becomes more difficult. There is no acknowledgment of a jobless recovery or of the intensified polarization of work opportunities. The result is a Kafkaesque buffet of policy "solutions" such as eliminating eligibility for UI if you voluntarily quit your job, or reducing access to social assistance if you don't take that job/training.

The vision of society that underlies these developments evokes a return to the dark ages of social policy, the pre-Depression period. There are no guarantees. The rights of citizenship have shifted from entitlements to the responsibilities of individuals and more limited "mutual" obligations between the state and individual. The message of the rugged individual is pervasive, with the poor being told they are responsible for themselves and the rich being assured they are responsible for no one *but* themselves. The terms of discussion have changed from creating a humane social minimum to the language of personal survival and getting by. This path returns us to pre-Depression notions of social security, but with a twist: in an atomized society, with less reliance on the extended families and communities of yesteryear, survival is increasingly based on one's own wits. We are becoming Margaret Thatcher's society of individuals.

How to change the course of this trajectory? Quite apart from long-term goals, can Canadians survive the immediate crisis? Can we afford to hang on to what we've got? While Canada's wealth (counted in dollars) has been increasing, its distribution has been growing more unequal. Remember the $5 billion that could have been freed up in 1991 alone if an already inequitable system of income distribution had not been permitted to grow more unequal? Reorganizing priorities within the budget, implementing a progressive tax system for both personal and corporate income, and "rooting" more of our Canadian-made savings at home would provide more than enough to cover the costs of

the current system of social security for the foreseeable future—
provided that the most dominant trend in the labour market
does not continue to be an increasing proportion of people outside the labour market.

That is the nub of the problem. Postwar society was built
around the twin pillars of jobs at living wages for all who sought
work and income supports for those who found themselves outside the labour market. But the security of income through the
labour market is deteriorating. With the economic system
increasingly prone to failure (from the perspective of the workforce), the predictable result is the mounting costs of mopping
up after its fall-out. The policy response has been to contain costs
through eroding the terms of entitlement to protection and service. If the one pillar of employment continues to disintegrate,
the building will slowly cave in. But if the opportunities offered
by the labour market are crumbling and the only "repair"
offered is to hack away at the second pillar, the system of income
security, the structure will surely collapse much more quickly.

This leads to the conclusion that the affordability of social
programs should not be the first focus of concern. Nor, given the
current political and economic climate, is it an efficient use of
resources or tactically smart to address the question of how to
"redefine" the social/income security system. Rather, we should
be pooling our imaginations, intelligence, and political strength
to ensure the availability of decently paid work. That challenge
has been swamped by the realities of half a decade of rearguard
fighting, of struggles to protect the hard-won rights of Canadian
life. Nonetheless, the fundamental challenge to the progressive
community in this war of words remains: to refocus public attention away from competition and cost containment onto the real
engine of development, both social and economic—meaningful
work for all.

The Challenge — Rerevolutionizing the Meaning of Social Security

It is hard to use the term "progressive" without flinching when so much of the left's recent effort and thinking has been devoted to defending the status quo, a status quo that not long ago was severely critiqued. Are there really no alternatives? The force of the backlash from the corporate agenda is so powerful that at times it seems market logic really is inevitable. Yet we know from history this is not so.

With economic growth on the horizon it becomes easier to imagine more humane ways of sharing the future. But the economies of the world now appear to be stalled for the foreseeable future. At best we seem headed for a stagnant and "jobless" economic recovery, at worst we are about to cross over the low-inflation trip wire into a deflationary spiral. People are being caught in a squeeze play between the low-or-no-growth reality and increased pressures to "reform" (read cut back) welfare, UI, and other forms of income security. The past decade has produced a real transition in the public's mindset, a rejection of the relevance of social forms of well-being towards a focus on what is best for personal salvation. The economic take on this scenario says the only path to salvation is to focus on the off-shore world (global capital, export orientation of production) and the inner world (training and adjustment). All these forces combine to undercut notions of collectivity. We are being conned into thinking that there is no role for the nation-state and that national campaigns to effect change have no impact.

What can be offered as some realistic first steps to shift tracks from this road of bankrupt policies that only leads to more of the same? First, it must be acknowledged that the public discussion of social security has been successfully hijacked. It is critically important to re-examine what social security could mean, to recast our thinking about its significance in our lives. The twin challenges for this undertaking involve: 1) developing a new language about full employment, based on domestic principles from both the perspective of how and what we produce, as well as tactics to capture and root the money required to do the job; and 2) creating a consensus about the importance of redistribution. I'll address the second challenge first. There is already a long dialogue about issues of full employment, but the thorny implications of redistribution are rarely tackled.

The Clarion Call of Redistribution

The most distinctive element that the progressive community can bring to this new war of ideas is a concrete set of proposals to force discussion about the need for redistribution. All indicators and studies and day-to-day experience point to the need to redress the inequities that continue to grow more obscene through boom times as well as slumps. The evidence is unmistakable: even economic growth can no longer breathe life into our dreams of a fairer society.

A necessary starting point is the development of a program of wage solidarity. This would require, first, raising minimum wages in every province (to 60 per cent of the average industrial wage). Second, it would mean establishing "ceilings" for earnings, which can be done through a variety of means. The tax system is the most potent weapon for achieving this type of equity, either corporate or personal. In the personal income tax system, some threshold of income would be chosen (for example, the average pretax take of the top decile, $125,000), above which earnings and returns on investment would be taxed at very high marginal rates (more than 50 per cent).[37] As for the corporate tax system, Minnesota Republican Martin Sabo provided an example of a possible approach in July 1991 when he introduced the

Income Disparities Act in Congress. The act mandated that companies paying executive salaries and bonuses of more than twenty-five times the salary of the lowest-paid employee (representing the difference between the U.S. president's salary and the federal minimum wage) would have the excess amount subject to the highest corporate rate (34 per cent).[38]

Redressing the imbalances in the distribution of market earnings, however important on the yardstick of social equity, does not begin to address the underlying unease in society arising from the unequal distribution of work. It is time to again raise the issue of redistributing working time, although the old thirty-for-forty recipe to reduce working hours but retain income will not be accommodated in a climate of cost-containing labour-shedding. The most attainable option at the moment would seem to be a renewed focus on creating statutory entitlements for leave, assuming that there could be effective limitations on the use of overtime. These entitlements would be backed up by social insurance systems that permit all working people (full-time and part-time) to cash in on such benefits by providing reasonable income replacement while on these leaves.

Creating a system in which these entitlements would be accessible to all workers indirectly addresses the question, "Is universality dead?" From some perspectives this principle does appear irretrievable, especially in terms of direct transfers of income. The momentum for continuing to target and rationalize systems of social assistance (welfare, child benefits) is bleeding through to forms of insurance (currently UI, soon pensions). But universal access to services is not out of reach and has wide public support. These would include: medical services; a real right to lifelong learning (through entitlements to education at all basic levels and creating mechanisms for easing access to higher technical and liberal arts learning); a nationwide system of universally accessible, high-quality child care; a nationwide system of supports to the elderly in their homes; and statutory extensions of benefits for part-time workers. In all these cases, universality remains the central underpinning, and the process of expanding access to such services and setting national standards is an important step towards ensuring better equity among Canadians.

Among these needs, the two most glaring vacuums in current policy are leaves for educational upgrading and parenting. Training has become every government's panacea for the ills of the labour market. Yet, despite a 30 per cent drop-out rate from secondary school and an unprecedented degree of structural change in the economy's skill mix, there is a large gap between needs and resources. Both training for the employed and government-sponsored training focus primarily on job-specific and short-term programs. In most cases the quick-fix approach cannot address the needs of people who need to upgrade their skills and knowledge base.

To make the statutory leave for educational upgrading a real option for all, either on an annual or a more cumulative "sabbatical" basis, people need access to income support. Workers could rely on unemployment insurance benefits (if the scheme were expanded for educational purposes, not just unemployed "trainees" in short-term programs), and those who do not qualify for UI could receive a national training allowance. The financing of the project could come from three sources: increased spending by government through the Consolidated Revenue Fund (the tax base); broader coverage by the UI system, implicating some increase of premiums at least in the short term; and a payroll tax on employers that currently benefit directly from government-subsidized on-the-job training programs. Ironically, this example of universal access to service or entitlement is predicated on a system of income support that itself is universally available.

The absence of a national strategy to deal with child care is unfathomable, given the explosion of the participation rate of women with preschool-age children, which stood at 67.2 per cent in 1991 (up from 50 per cent in 1981 and 35 per cent in 1971) and continues to rise. This strategy should be twofold: first, establishing a universally accessible, high-quality system of day care, to be as available in rural centres as in urban centres; second, the provision of income support so that the option exists for children, regardless of their socio-economic class, to be cared for by their own parent(s) in the first two to three years of their lives (that is, until the stage at which children are usually able to begin to communicate for themselves). In Europe supplementary child benefits and/or family allowances are highest for children up to three years of age. These payments could be financed either by extending UI benefits and/or enhancing child benefits (since family allowances no longer exist). The tax system can be our partner, again, in ensuring that there are no perverse, unintended effects, such as paying the spouses of wealthy earners to stay at home. Again, both the service (day care) and the income support would be universally available to parents. The goal is to restore choice in the way people construct their lives.

It would be ideal to campaign for reduced working hours, especially overtime, but the pragmatic question is how to enforce such measures? Two phenomena would tend to swamp this effort: first, the rapid expansion, in every sector, of a just-in-time labour force, in which the effort to limit overtime misses the point about how production is being organized with an expectation of workers being on-call twenty-four-hours-a-day for when they are needed. They get no work when they are not needed. Workers, and in many cases employers, simply cannot say no to extended hours. Second, many workers want the long hours in order to make a decent living. It will be difficult to convince them to voluntarily cut their income, especially if they are working at very low wages. The dilemma is that, as the argument for more leave time develops, relatively well-off workers with provisions for paid leave and decent vacations are not taking what for decades they have gained through collective bargaining. This tendency endangers the original rationale for these things: the need for rest, relaxation, and family time. It is time to challenge the reasons why workers and employers seemingly can't say no.

A campaign that places the premium on redistribution in a climate of economic stagnation or decline is admittedly a risky proposition. The value of pursuing it, win or lose, is to enlarge the scope of public discussion about the choices facing society at a time when people are beginning to seriously believe there are no alternatives to further cutbacks.

Reframing the Terms of Full Employment — Investment and Domesticity

Redistribution itself cannot solve the most critical dilemma: how to create enough decently paid work? To reach this goal we must regain some degree of control over capital mobility. Yet policies to this effect fly in the face of current developments, whose overriding thrust is to eliminate barriers to the free movement of capital.

Understandably, this is the most politically, and perhaps economically, prickly problem. Yet it is not as if there is no room to move, that all the money suddenly evaporates if changes start to occur in how it is used. While global capital has become more of a reality in how activities are financed, there is still a huge pool of domestic capital being generated and saved every day by the people who live and work in Canada. For example, as of December 1990, the total (book value) assets of Canadian trusteed pension plans had reached $199.783 billion. (Compare this to the total Canadian dollar assets of the chartered banks at the same time of $405.293 billion.) In February 1990 the federal government changed the limits on foreign investment of these funds, raising the ceiling from 10 per cent to 20 per cent by 1994.[39] A simple change back to the old regulations would "root" almost $20 billion back for investment in Canada. In the words of an ad campaign for beer, "Change the rules, and you change the way you think about things."

Think about tax expenditures on Registered Retirement Savings Plans (RRSPs), an enormous leak in the system's ability to capture revenue that could be easily plugged by changing threshold levels of exemptions. In December 1992 the Department of Finance released a small document, "Government of Canada Personal Income Tax Expenditures," which showed that in 1989 alone $4.8 billion in tax revenues were forfeited due to terms governing deductions for contributions and the non-taxation of investment income earned inside an RRSP. Similar figures for registered pension plans total a further $10.6 billion. Think of the 1992 Auditor General's report, which found that in 1990 "Canadian corporate taxpayers 'invested' $16.1 billion in . . . tax havens. . . . It is reasonable to conclude that hundreds of millions of dollars in tax revenue have already been lost and will continue to be at risk. . . . It is important that the reviews of interest deductibility, foreign source income and foreign affiliates previously announced be completed."[40] Think about why the Progressive Conservative finance minister Mazankowski publicly rejected these suggestions the day after the report.

The need to attract global capital is evoked again in connection to our debt. Canadians used to finance their debt almost entirely through Canada Savings Bonds, a long-term financial instrument that can be held only by residents. Today most of the Canadian debt is serviced by short-term instruments such as Treasury Bills, and the level of foreign ownership of Canadian unmatured debt has been increasing steadily, standing at 23.6 per cent by the end of the 1991-92 fiscal year compared to 8.9 per cent in 1983-84.[41] Continued deficits will bring us to a "debt wall," say the pundits, with foreign sources of capital refusing to provide financing—though there is not a financier on the surface of the planet who would refuse a higher rate of interest to do so. What would be the feasibility, costs, and benefits of reintroducing an explicit policy of servicing Canada's debt at home?

Similarly, the economic theology of the day values the flowing dollar and discounts the dollar that stays at home. In the name of prosperity, it has become a God-given right to buy German marks or Mexican pesos. The focus on the global marketplace, and on finding niches for our Canadian stars, makes it alright to invest outside the country, where the dollar buys nothing at all for Canadian citizens; but the dollar given to the welfare recipient or the unemployed worker who spends it locally on rent, public transport, or groceries is seen as going down the drain. This type of consumption is begrudged by policy-makers and Canadians who perceive themselves foremost as beleaguered taxpayers, not as people who could someday be exposed to the risk of needing to rely on UI or even social assistance.

Ironically, although this model is promoted as the only way to do things, an opposite model is emphasized by white policy-makers negotiating with Indian bands: root some of the money from government into the reserve. Native people are told that the only way to promote a greater degree of economic health is to slow the money hemorrhage by investing locally, building up local businesses, and creating their own products. Somehow our economic collectivity and theirs need to operate on different principles.

Discussions of how investment works for Canadians at home do not go far without turning to matters of trade. Here the language of managed and fair trade, sectoral agreements, and procurement policies is developing. But there are also existing examples of how these systems could work. The auto pact was arguably the engine of Canadian industrial growth during the 1960s and 1970s. How reproducible is this type of content legislation in other sectors where there are many, rather than few, producers, such as the clothing, food production, and electronics industries? What are the lessons to be learned from supply management in agricultural products and non-tariff controls through quotas? Careful examination of these and other arrangements will provide ample ideas for future experimentation; but the promise and limitations of these ideas can only be realistically assessed when they are put into practice.

Where these ideas get tested first is also critical. These considerations urgently need to be framed within a public discussion about what sectors are of national interest. The sectors should be defined in a manner that really does include all citizens: all Canadians need food, clothing, shelter, and water. To this list one can add energy, furniture, appliances, or pharmaceuticals. The basics of life, rather than high-value-added, niche-oriented products and services, should guide the choice of sectors in which Canadians, as well as people in other nations, demand the basic right of being able to produce a significant proportion of what they need to consume.

The global marketplace, as irrevocably important as it has become to all interdependent economies, tends to undercut this simple premise. The orientation I suugest is blatantly domestic, in every sense of the word; but it is also global in its applicability and appreciation of how production is now being practised internationally. If people of different nations were to demand such an approach of the governments that supposedly represent their interests, it may be possible to break the vicious cycle of global competition that pits us, as consumers, against ourselves as producers. This could be the necessary precondition for developing a bulwark against the endless international trap of competing to attract investment at any cost.

National governments still clearly play a pivotal role. In fact, that is the only meaningful level at which the issues of growing inequality can be challenged and addressed through the development of policies of redistribution. National governments should be expected to carry the voice of their people to business and to the international community by setting limits on the culture of "passive consumption," demanding that the marketplace must provide a living as well as a lifestyle. Significantly, this entails reregulating aspects of the economy, which in turn requires breaking out of the lock-step of international trends in governing. The key question is: does the first "leader" to try a new direction pay the highest cost? Can Canada provide examples of different approaches to these problems—examples that could be emulated in other parts of the world? We have done so before. Would even small changes to the regulatory climate provoke a capital "strike"? If so, how much of this type of economic blackmail would cripple our economy? These are empirical questions. Just as we have the empirical results of the test question "what are the costs and benefits of deregulating the economy?" so we will not know the costs and benefits of such changes until we try them.

Conclusion

Since the 1930s Canadians and their elected governments have pushed and shoved demands and policies back and forth, shaping and defining our version of social security in different ways over time. Now, in the 1990s we are entering a critical stage of this dialogue, with the distinct possibility of moving closer to a pre-Depression model of insecurity. What are the current terms of the debate?

There are three basic arguments: 1) our system of programs is out of date and does not serve the needs and interests of people in a rapidly changing and globally competitive economy; 2) our system of programs promotes dependency and should be reformed to develop self-sufficiency and enhance employability; and 3) we can't afford these programs, anyway.

The first two points are transparently self-serving rhetoric plucked from the corporate lexicon. It is the very volatility of the globally competitive economy that has made income support programs more essential than ever, leading to the perverse situation of programs being cut back because they are so necessary. Overhauling social security emerges as a politically expedient move, something that makes governments look like they are doing something about the crisis; but the crisis is about economies facing possible system breakdown and about the significance of nation-states. If we do not break out of deficit logic, then we can only continue to reorganize the deck chairs on the Titanic.

A war is being waged to get people to accept that 1) there is no option to heightened competition for fewer and/or worse jobs, and 2) the role of a national government has become largely irrelevant to the individual. But there is absolutely no immutable economic logic that says this must be so, as history has abundantly shown. Governments can and do act as arbiters between the market and people, and there is no reason why they cannot continue to do so. To accomplish this task the most obvious level, and still the most powerful, remains the nation-state.

Possibly because the crisis does not appear to have an end in sight, people do entertain the legitimacy of the third point, "Yes we need it, but there's no money left to do these things." While the evidence in this chapter shows that indeed the bills can be paid in the medium term, it is also true that if nothing is done about the economy the system of income security will eventually become unaffordable. And this is precisely the launching point for the counterassault in this war of ideas.

We should be breaking down the constrictions put on this debate, reframing the whole problem. Debate about welfare reform would not be getting so much air time if the economic sphere were functioning properly, namely providing decent jobs. But calls for full employment are not enough. This is the moment to be establishing principles for future social and economic security: a more equitable distribution of time and money, the right to produce a meaningful proportion of the goods and services we need to consume, and the opportunity to disengage from the market when our family or personal needs dictate. This means more jobs, but not jobs necessarily tied to a model of economic growth or chained to the caprices of business cycles. It also means defending and expanding the system of supports we now have.

The logic that says further shrinkage is necessary and inevitable is deeply flawed. The path of continued belt-tightening will never provide enough jobs for everyone. It will constantly put the safety net under threat of further erosion and, ironically, will never pay off the debt.

If competitiveness by cost-containment becomes the scale by which every decision in public policy is measured, the deliberations can only be about what levels of guaranteed unemployment will be tolerated and what degree of guaranteed poverty is politically salable. This leads irrevocably to the tragedy of our times—the move to take apart programs set up precisely because of the experience of the Depression, just at the time when that experience is being re-created by an unswerving allegiance to deficit logic, unregulated markets, and the erosion of the rights of citizenship. More fundamentally, that tragedy diverts our attention from the obvious social equalizer: full employment, founded on a common understanding that work for all is important for both social and economic reasons.

Notes

1. Edward Ng, "Children and Elderly People: Sharing Public Income Resources," Statistics Canada, Catalogue 11-008E, *Canadian Social Trends*, Summer 1992, p.13; and Statistics Canada, Catalogue 13-207, *Income Distributions by Size in Canada*, 1991.

2. Unpublished data, Daily Bread Food Bank annual surveys and the Canadian Association of Food Banks.

3. Calculated from Statistics Canada, Catalogue 71-001, *The Labour Force*, annual average data from historical series (starting 1946) and various years.

4. In the end the federal government's weak commitment to the institutionalization of full employment was never seriously tested, as an unforeseen period of economic expansion started almost immediately. See Leslie A. Pal, "Tools for the Job: Canada's Evolution from Public Works to Mandated Employment," in *Canadian Welfare State: Evolution and Transition*, ed. Jacqueline Ismael (Edmonton: University of Alberta Press, 1987).

5. See Donald Smiley, ed., *The Rowell-Sirois Report*, 1963, pp.185-87. This was the 1937 Report of the Royal Commission on Dominion-Provincial Relations, the first effort to create a structure of "equalization" payments between federal and provincial governments and between have and have-not provinces. The goal of such an exercise was made explicit: The Commission "attempted in its fiscal proposal to provide both for a more equitable distribution of governmental burdens and social service benefits throughout Canada and to make possible a revenue system and a general fiscal policy designed to stimulate rather than depress the national income" (emphasis added). See Book 2, p.79 of the report. By 1957 the series of fiscal arrangements that were made throughout the war were codified in formalized equalization grants, payments based on meeting the national average of revenue raising capacity.

6. Section 36 of the Canadian Constitution. These clauses in the 1982 repatriation of the British North America Act to Canada specifically refer to equalization.

7. *Social Security for Canada*, report to the [federal] Advisory Committee on Reconstruction, February 1943, p.107. This is also known as the Marsh Blueprint after the research advisor of the report, Leonard C. Marsh.

8. Frank Strain and Derek Hum, "Canadian Federalism and the Welfare State" in *Canadian Welfare State*, ed. Ismael, pp.365-367. The authors note that the federal Old Age Pension Act of 1927 was the first shift in the notion of citizenship. Although administration was through the provinces, these plans had to conform through federal approval with national standards, and a cost-shared agreement shifted part of the financial responsibility to the national society.

9. *Social Security in Canada*, p.18.

10. "Without an investment mobilization against the first shocks of economic

demobilization, without an employment programme to sustain the country while industry and agriculture are moving towards a new peacetime equibilrium, the social insurances structure will have no solid foundation." Ibid., p.40.

11. Note that the jump in unemployment was not caused by a recession. In fact the economy was still growing in real terms at an annual rate of between 2 and 5 per cent throughout this period.

12. Greg Albo, "The Great Automation Scare and Canadian Unemployment," unpublished paper, May 1992.

13. Pal, "Tools for the Job," p.49.

14. Derek Hum, *Federalism and the Poor: A Review of the Canada Assistance Plan* (Toronto: Ontario Economic Council, 1983) p.19.

15. See *The National Finances: An Analysis of the Revenues and Expenditures of the Government of Canada, 1988-89* (Toronto: Canadian Tax Foundation, 1990), Table 3.7.

16. The three main reasons given are: 1) the reduction in effective corporate income tax rates; 2) proliferation of tax expenditures in the personal income tax base (which narrowed the tax base from 1972 to the mid-1980s); and 3) reductions in the effective sales tax rate — at least until the introduction of the GST. See Irwin Gillespie, *Tax, Borrow and Spend: Financing Federal Spending in Canada, 1867-1990* (Ottawa: Carleton University Press, 1991), pp. 211-17.

17. Andrew Johnson, "Ideology and Income Supplementation," in *Canadian Welfare State*, ed. Ismael, p.81.

18. Mario Iacobacci and Mario Seccareccia, "Full Empoyment versus Income Maintenance," *Studies in Political Economy*, No.28 (Spring 1989), p.149.

19. Just as there is high degree of turnover in the marginal labour market, so too is there more fluidity in and out of poverty, at least according to data based on the period 1982-86. See Economic Council of Canada, *The New Face of Poverty* (Ottawa, 1992), pp.22-25. For recent data on how long people stay on assistance during periods of crisis and/or joblessness, see A. Yalnizyan, "Market Madness," *Social Infopac*, Vol. 12, No.1 (Social Planning Council of Metropolitan Toronto, February 1993), p.2.

20. Martin Rein, quoted in Patricia Evans and Eilene McIntyre, "Welfare, Work Incentives, and the Single Mother," in *Canadian Welfare State*, ed. Ismael, p.110.

21. *The National Finances*, Table 3.7.

22. Here we define social spending as transfer payments (including unemployment insurance, old age security and guaranteed income supplements for the elderly, family allowances, and veterans' benefits); insurance and medical care, the Canada Assistance Plan, and educational support. These figures are calculated from Table 1.

23. In 1951 corporate taxes were a greater source of revenue (30.6 per cent) than

personal income tax (26.1 per cent). By 1991 54 per cent of federal revenues were generated from personal income taxes, while corporate taxes accounted for only 11 per cent.

24. Thomas J. Courchene, *Social Policy in the 1990s* (Toronto: C.D. Howe Institute, 1987), p.179.

25. Grattan Gray, "Social Policy by Stealth," *Policy Options Politique*, March 1990.

26. Ken Battle, Caledon Institute, May 1992. Both the tax system and child benefits have been partially de-indexed, giving the government a "perpetual-motion money machine" so long as there is inflation.

27. The 1984-85 CRF allocation for training was $2.1 billion. The estimate for 1992-93 was $1.6 billion. Adjusting for inflation this represents a decline of 45.8 per cent.

28. The counterpoint to this theme is the attitude of people living in countries such as Germany, France, and Sweden. In Europe it is commonly understood that the popularity of and support for social security programs are directly defined by who gets them; the more you target them the more they are reviled.

29. Milton Friedman developed a basic income approach using the tax system as early as 1958. Although there have been many different versions of a guaranteed income, the idea still has the cachet of being promoted by the left. But the specific policy designs proposed more recently have come straight out of the minimalist and residual school of basic incomes rather than the more progressive version based on combatting poverty by raising the social minimum. Our system of social insurance and assistance programs already provides a guaranteed income to virtually every citizen in Canada, albeit categorically. The problem with the "all-purpose" basic income approach that surfaced in the 1980s is that the construction of a single "floor" of income in society opens the door to attacks on the minimum wage and programs such as UI. (See *Report of the Royal Commission on the Economic Union and Development Prospects for Canada*, Vol. Two, Part Five, pp.619-22.) The lower the floor, the greater the competition among and between workers and the unemployed, especially at the bottom end of the income spectrum. Rather than alleviating poverty, this form of guaranteed income reinforces and exacerbates poverty.

30. Economic Council of Canada, *Good Jobs, Bad Jobs* (Ottawa: Supply and Services, 1990); and Armine Yalnizyan, *Reflections on Full Employment* (Toronto: Social Planning Council of Metropolitan Toronto, 1991).

31. See "The Jobless Recovery" Five-part series in *The Globe and Mail*, Report on Business, Jan. 11-15, 1993 (various authors); and John Greenwald, "The Job Freeze," *Time*, Feb. 1, 1993, pp. 34-35. For an analysis of the growing magnitude of this phenomenon in Toronto, see the discussion in Yalnizyan, "Market Madness."

32. This discussion of the SSP was taken from "Design of the Self-Sufficiency Project," prepared for Employment and Immigration Canada by Social Research Demonstration Corporation in co-operation with Manpower Demonstration Research Corporation, June 24, 1992.

33. The document "Design of the Self-Sufficiency Project" raises questions about how voluntary participation will be. After having identified people who fit the criteria of the target group, "interviewers will contact each month a random subset of these eligibles ... and obtain release of information and informed consent from clients. Individuals will ideally be contacted by telephone and then interviewed in-person at their homes." Should the person be selected to be in the program, rather than in the control group, they "will be asked to enroll in the program and attend an orientation. ... Those who did not attend will be contacted again" (p.24).

34. The upper threshold of earnings against which the supplement is calculated is called a target income. In British Columbia the target is $35,000, in New Brunswick $25,000.

35. Benoît Bouchard, "Canadian Paper on New Orientations for Social Policy," speech delivered by the Minister of National Health and Welfare to the meeting of the Employment, Labour and Social Affairs Committee at the Ministerial Level on Social Policy, Paris, France, Dec. 8-9, 1992.

36. Bouchard, "Canadian Paper on New Orientations for Social Policy," p. 10. Emphasis in original.

37. A marginal tax rate of 100 per cent would send out a more serious message, not only about the earnings gap amongst citizens of a nation, but what level of affluence is tolerable in a global setting with environmental limits to consumption. Some would respond that a 100 per cent rate would simply be inflammatory posturing. But this is the same marginal tax rate that is applied to certain categories of earnings for welfare recipients.

38. "How Much Dough for the Big Cheese?" *Utne Reader*, March/April 1992, p.18.

39. Information on book value of trusteed pension funds is from Statistics Canada, Catalogue 74-201, *Trusteed Pension Funds*; on changes in government-set limits on foreign investment of these funds from *Survey of Pension Plans in Canada* (Ottawa: Financial Executives Institute in Canada, 9th edition, 1992), p.80; on total Canadian dollar assets of the chartered banks from *Bank of Canada Review*.

40. *Report of the Auditor General of Canada to the House of Commons, 1992* (Ottawa: Supply and Services, 1992), pp.50-51.

41. About 90 per cent of Canada's accumulated deficit is held in the form of unmatured debt. The proportion of foreign-held debt is calculated from figures provided in the (monthly) *Bank of Canada Review*.

The New Tools: Implications For The Future of Work

*T. Ran Ide and Arthur J. Cordell**

> Without work the vessel of life has no ballast.
> — Stendhal (1788-1842)

In the early 1960s, writing about the world's electronic inter-dependence occasioned by the rapid growth of twentieth-century mass media, Marshall McLuhan introduced the concept of the *global village*. Some thirty years later it is fashionable to use the term *globalization* in discussing competition among the free market economies. In the Reagan/Thatcher era the apparent success of hard-nosed capitalism in providing a plethora of consumer goods, together with the growing economic power of Western Germany and Japan, convinced many authorities that nirvana depended upon the adoption of the free market system.

While the collapse of both the Soviet Union and the political

*The authors wish to thank Betty Gerow and Arlene Ide for their valuable editorial advice, Bill Graham for technical assistance, Frances Bonney and Andrew Reddick for their research help, and Brigitte Boucher for her unfailing enthusiasm.

structures of other member states of the Warsaw Pact had a number of underlying causes, there is no doubt that televised images of luxury-laden shelves contributed to popular unrest in those countries. These events seem to be a case of the global village and the globalized market writ large.

The decade of the 1990s began with what was described as a mild recession. In many countries where the so-called free market economy prevailed, unemployment rates rose even as productivity increased. The reasons for this anomaly have either not been studied at all or studied only at the most superficial level. Conventional wisdom has attributed much of the cause of the recession and resulting unemployment to the huge national debts amassed by the United States and other industrialized nations, together with the policy of increasing annual deficits to avoid raising taxes. Added to these explanations are the larger than usual corporate and consumer debts that come from the growing expansion of credit. Whether or not we are facing a modern "South Sea Bubble," or a much more serious international financial crisis, the signs are ominous.

Ideally the 1990s ought to be the best of times.[1] The new technologies, particularly those related to microelectronics, would seem likely to bring important benefits. They are clean, highly efficient, and productive. The goods they turn out are more cheaply made and require little of the monotonous and backbreaking labour so characteristic of the past. But these are not the best of times. There has been a slip somewhere between the promise and the performance. The nature of work as we used to know it has changed.

The results are a greater disparity of income and quality of life between the haves and have-nots, the erosion of the middle class, and, above all, major increases in unemployment in most developed nations. Even a conservative publication like *Business Week*, in discussing downward mobility, has conceded that a recovery, if and when it comes, will not bring back the thousands of jobs lost by managers and professionals.[2] The consequence has been a malaise that has undermined the confidence of people in their political, academic, and economic institutions.

Is it the new technologies that are at fault? Should we blame the computers, the robots, digital transmission systems, et cetera? After all, they are merely tools, more complicated perhaps than shovels, axes, and screwdrivers, but tools nevertheless. Tools are inanimate objects. They do not make moral judgements — people do. But tools can be used in different ways, some constructive and others destructive. Apart from such obvious examples as weapons, the designers of tools usually see them as instruments to alleviate labour and/or improve the quality of life. In *Tools For Conviviality* Ivan Illich added another dimension, warning, "As the power of machines increases, the role of persons more and more decreases to that of mere consumers."[3] While he wrote the book some twenty years ago, the concern expressed still seems relevant today, during a time when more and more products are designed to meet the most transitory of needs. Of course, Illich's message went far beyond this generalization. He believed that convivial tools were the ones that provided the users with unique opportunities to express themselves as well as to enrich the environment with the fruits of their vision and labour. He concluded: "Defence of conviviality is possible only if undertaken by people with tools they control." Although some critics may consider the Illich position to be impractical, we would argue that in discussing the future of work and the implications of the technological revolution there is a need to recognize the dangers inherent in a retreat from the humanitarian and liberal ideals that accompanied the introduction of democratic systems.

The History of Work as Employment

The word "work" is used in many senses. A physicist might speak of it as the product of a force applied to a body. A mathematician might be referring to an operation in calculation, or an artist might be talking about a painting or a musical or literary composition. The term *working class* usually refers to those members of society who are employed for wages in industry, and workers are often equated with *labour* as distinct from *management*. It is important, therefore, to define what we mean by work.

In the concluding chapter of *Microelectronics and Society*, Adam Schaff distinguishes between *work* and *occupation*, reserving work for activities related to the production of goods or services and applying occupation to how people spend their time.[4] In common parlance work is most often thought of as what one does to earn a living, and this is the sense that we use here.

The economic problems society now faces relate in part to the changing nature of work. In the recession of the 1990s, jobs are being changed or lost altogether. Some people relate the cause to cyclical fluctuations that permit economies to adjust to changing conditions, so that after a certain period of time normalcy returns. Others attribute the difficulties to the so-called global economy and the inability of the traditional industrial nations to compete with countries where labour costs are low or where new technologies have been developed unhindered by large capital investments in obsolete equipment.

If it were only so simple — then financial houses could be put in order, deficits reduced, and old methods of production could be replaced by new ones. But even if this were to happen, a return to the sense of affluence felt in the 1980s would be unlikely without major changes in the value systems that characterize the era of the free market. These are the same value systems that in toting up the numbers to arrive at a "bottom line" have underestimated the importance of the human being.

Over recorded history the ways in which men and women have struggled to survive in a largely unfriendly environment have changed radically. In earlier periods our ancestors were forced to hunt or fish for their food and find shelter where they could. They went through a pastoral stage, an agricultural stage, and a handicraft stage that marked the beginnings of specialization. In the eighteenth century the invention of power engines to run spinning and weaving machines led to what became known as the Industrial Revolution. Human lives were transformed. Entrepreneurs took over from landowners; farmers still farmed and fishers still fished; but people moved in droves from rural to urban areas, where the factories were located. Once started, the rapid stream of innovative technologies has never stopped.

Adam Smith may not have been the founder of political economy, but with the publication of *The Wealth of Nations* in 1776 he became its pre-eminent spokesman in those early days. To illustrate how wealth could be increased by an efficient use of the labour force, Smith used the example of the manufacture of pins. Instead of having one worker carry out all the tasks of making a pin—drawing the wire, cutting, head-fitting, and sharpening—the firm should have each worker specializing in one operation only. That way, Smith pointed out, instead of a minimal output, production would be increased a hundredfold.

After Smith, David Ricardo, John Stuart Mill, and Thomas Malthus became recognized authorities as England spearheaded the drive towards industrialization and as "land, labour, and capital" became identified as the factors of production.

In looking at production and industry and assessing their contributions to a nation's economy, we often use the terms *goods* and *services*. However, the distinction between the two is becoming less relevant as more of the value of goods is increasingly attributed to the services entailed in the product. Robert Reich, speaking about steelmaking, points out: "When a new alloy is molded to a specific weight and tolerance, services account for a significant part of the value of the resulting product. Steel service centres help customers choose the steels and alloys they need, and then inspect, slit, coat, store and deliver the materials."[5]

Keeping in mind that it is increasingly difficult to distinguish between the two, it is still helpful to look at goods and services as a means of determining what is happening to employment and, in the larger sense, work. Before the Industrial Revolution, the majority of people found their livelihoods in primary occupations such as agriculture, mining, and fishing. But the percentage of the workforce engaged in those activities steadily declined, and still continues to do so. At the same time the numbers of people engaged in manufacturing dramatically increased—at least until relatively recently, when more and more new jobs have been created in the service sector.

Much of the reason for this change relates to automation. Annual productivity increases in agriculture were minuscule in the early years of the nineteenth century, but productivity began to increase by leaps and bounds with the introduction of new tools such as hay loaders, tractors, harvesters, cream separators, milking machines, and farm machinery of all kinds. As a result agriculture required less labour, but more capital. Farms became larger and more highly specialized, and many agricultural workers were forced to find employment elsewhere.

Fortunately there was a place for them to go. Factories were sprouting up everywhere there was reasonable access to raw materials and an adequate transportation system. The demand for labour, both skilled and unskilled, was high. Towns and cities became larger as the need for housing of all kinds increased. A mass transformation in the way people earned their livelihood was taking place. Confounding the theory of Thomas Malthus that population growth would be checked by poverty or some other calamity, the number of people continued to increase, and they seemed to be better off than ever before.

But while a few people may indeed have been doing better, most workers were not. Little attention was paid to sanitation, and houses were filthy. Factories were damp and poorly ventilated, women and children were exploited, and the hours of work often exceeded twelve a day. As time went on things began to get better. Trade unions forced improvements in wages and working conditions, and social critics argued for a just payment for labour, touching the conscience of the enlightened public.

One of these social critics, John Ruskin, wrote, "The universal and constant action of justice is . . . to diminish the power of wealth, in the hands of one individual, over masses of men, and to distribute it through a chain of men."[6] This statement became one of the first arguments in favour of the equitable distribution of wealth. In 1860 Ruskin's book *Unto This Last*, which attacked the conventional wisdom of the day, was published following several attempts to have it banned. After a slow start, by the turn of the century it had sold over one hundred thousand copies and become a significant agent of change. Much later on Gandhi said the book had captured him and transformed his life. Tolstoy called him a person who thinks with his heart and one of the most remarkable men of all time. In his introduction to Ruskin's *The Bible of Amiens* Proust declared, "He will teach me, for, is he not he too, in some degree the Truth." The principle that there were moral conditions attached to work and wealth itself had been recognized.

Standards of living did improve and, except for the periods covered by the depression of the early 1930s and the two great wars in the industrialized nations, the age of affluence seemed to be at hand. Material goods were produced in plenty, universal education became the rule, and universal health care seemed to be a viable objective. In the United States of the 1960s President

Lyndon Johnson implemented a massive anti-poverty program and proposed to build a "Great Society" until his plans were interrupted by the Vietnam War.

But there was a clock ticking, counting down to what was to become another economic revolution leading to what would be called "The Information Society." It began, innocently enough, with an increasing interest in computation. In 1642 French mathematician Blaise Pascal, at the age of nineteen, built the first digital computer. Almost two hundred years later an English mathematician, Charles Babbage, designed a calculating machine that, although never completed, earned him almost universal recognition as the father of the modern computer. The combination of the use of a binary mathematical system with an algebra devised by George Boole in the second half of the nineteenth century (whereby the relationship between a large number of elements could be dealt with by repeated applications between a series of two elements) paved the way for the development of small solid state transistors and integrated circuits and finally of highly sophisticated computers. By 1971, when Intel Corp. in the United States introduced the first microprocessor, the age of microelectronics had arrived.

And what an age it has proved to be! The workplace has been transformed. Computer-assisted design (CAD), computer-assisted manufacturing (CAM), and computer-aided systems engineering (CASE) have radically altered how existing products are improved and new products created. No longer are long hours needed to examine and reject flawed designs: the computer can do this in seconds. No longer does Adam Smith's analogy of the manufacture of pins apply, nor do the production-line techniques devised by Henry Ford and others. Robots now perform the tasks in a fraction of the time and virtually free from error. The huge plant at the foot of Mount Fuji operated by Fanuc, the world's largest manufacturer of robots, turns out some twelve thousand machines annually, with a staff of less than ninety workers. The new production processes are both integrated and automated, and the workers formerly employed are supposedly free to move into other lines of work.

Economies of scale have permitted the development and marketing of a seemingly infinite number of new products to a degree never previously imagined, and at prices low enough to appeal to the widest range of consumers. Labour-saving devices such as self-operating washers and dryers for clothes and dishes, microwave ovens, food processors, water conditioners, and automated garage-door openers are common in the nations of the North. Along with television sets have come advanced home entertainment systems, including video-cassette recorders and compact discs to use in conjunction with multispeaker stereophonic playback consols. There are few households that cannot boast of more than one telephone, each with a variety of features that would have astounded the previous generation. Some phones are cordless, others store and recall selected numbers, and some have intercom capability. Some can be used to identify callers before the phone is picked up, and some have all of these features.

The marriage of computers and communications became inevitable with the adoption of digital transmission systems. As a result, the collection, storage, retrieval, processing, and production of information have been greatly improved. The computer led to word processors, video text, switched networks, communication satellites, and fibre-optic transmissions of voice, picture, and data in quantities and speeds unheard of in the past. These innovations have in turn resulted in improved air-traffic control, computer-aided instruction, and facsimile transmission of individual correspondence.[7]

As society moved from a period when labour was essentially manual to steam power and horsepower and now to computer power, the means of earning a living changed significantly. In the late 1880s relatively few people worked in manufacturing, but within decades their numbers had rapidly increased. A report by the Economic Council of Canada estimated that in the late 1940s, some 60 per cent of the Canadian labour force worked in natural resources, manufacturing, or construction. During the 1980s virtually all of the net job creation in Canada took place in the service sector, and by 1990 over 70 per cent were employed in services.[8]

Obviously, those employees able to do so found jobs in the rapidly expanding service sector. But what kind of jobs were these, and how secure will they be when the ubiquitous computer turns its attention to services?

To answer these questions and put the issue into perspective we need to look at the nature of wealth. Like work it has many meanings. Is it, as the *Oxford Universal Dictionary* defines it, 1) "the condition of being happy and prosperous" or 2) "the abundance of possessions." Or is it something else? We also need to look at the impacts of the industrial and information revolutions and try to see how they have influenced the quality of life. There is also the question of the nature of software and network intelligence as keys to an automated sector, and a need to better understand how they have affected not only individuals at the lower end of the income scale but also senior and middle managers. Only then can we hope to glean some perception of the magnitude of the problems currently facing society.

The Nature of Wealth

In the opening lines of *The Wealth of Nations*, Adam Smith identifies labour as the focal point of wealth: "The annual labour of every nation is the fund which originally supplies it with all the necessities and conveniences of life which it annually consumes and which consist always either in the immediate produce of that labour, or in what is purchased with that produce of that labour from other nations."

No matter that Smith goes on to talk about capital, interest, money, stock, rent, and the progress of opulence. Labour is central. Other theorists define wealth as being all useful things owned by people, but go to some lengths to limit those things to material items. Alfred Marshall, one of the leading economists of his day, noted in his *Principles of Economics* (1920) that wealth consisted of things that satisfy human wants—material goods that are transferable and immaterial goods that serve as a means of acquiring material goods.

Like most other theorists, Marshall excluded services that "pass out of existence at the same instant they enter it." He excluded values such as moral excellence, integrity, and honesty, as well as inherent or acquired skills. The genius of the artist, the inventor, or the theorist is not counted as wealth; that is, not until this ability has succeeded in producing "something useful." The only place for an Einstein would seem to be in his early trade as a patent clerk.

The Securities and Exchange Commission of the United States conducted a 1971 study in an attempt to determine the size of the national wealth. The authors considered wealth to be composed of reproducible assets, non-reproducible assets, and net foreign assets. Reproducible assets consisted of all types of buildings, producer durables, consumer durables, business inventories, and monetary metals. Non-reproducible assets were made up of various types of land holdings. Services as such were not accounted for. But surely wealth is more than purely material things, at least today, when slightly more than 70 per cent of the workforce is employed in the service sector. Regardless of how wealth is defined, these individuals must produce some things apart from the merely tangible. Today's discussion about sustainable development, accompanied by concerns about global warming and the fragility of the ozone layer, suggests that the nature of wealth is much more complex.

John Ruskin had a theory that said the only wealth is life. He arrived at it through a line of reasoning in which he maintained that the real test of production is "the manner and issue of consumption" and "the real question for the nation is not how much labour it employs, but how much life it produces." He concluded: "As consumption is the end of production, so life is the end and aim of consumption. . . . THERE IS NO WEALTH BUT LIFE. Life including all its powers of love, joy and of admiration. That country is richest which nourishes the greatest number of noble and happy human beings; that man is richest who, having perfected the functions of his own life to the utmost, has also the widest helpful influence, both personal and by means of his possessions, over the lives of others."[9]

Ruskin wrote this passage in 1860, over 130 years ago, but it is still a revolutionary idea. Today's discussions of wealth are still based on the ability to produce and exchange material goods. Real per capita income, that is, the gross national product divided by the population, is still considered to be the key indicator of the real wealth of a country.

There is another, perhaps latent, view of wealth, which is most apparent in discussions of threats to the environment. In effect, some important elements of society are saying, "Stop! There are trade-offs to be made." In this view there is an asymmetrical balance sheet. If we burn fossil fuels, continue to use aerosol propellants, produce chlorofluorocarbons for refrigerators and air-conditioners, and drill for oil in the Arctic, we may be creating jobs and wealth in the traditional sense. But the damage done by adding carbon dioxide and other industrial gases to the atmosphere, by depleting the ozone layer, or even by one oil tanker spill like the Exxon-Valdez incident off the coast of Alaska threatens our survival as surely as would the use of the nuclear weapons stored in the name of the defence of civilization.

While environmentalists have attempted to develop shadow prices to put these trade-offs into perspective, they still discuss damage to nature using the language of those who have for centuries dominated the discussion of wealth. It is impossible to put a price on the cost of upsetting the balance of nature, which carries with it the extinction of so many species of wildlife and the destruction of major forest areas. It is impossible to put a price on the loss of being able to walk in a wood or swim in unpolluted waters. It is impossible to put a price on the loss of pleasure and enjoyment that will be the inevitable consequence of unrestrained materialism.

It is also impossible to assess the human distress occasioned by the loss of work or the fear of loss of work. If, as Stendhal said, the vessel of life has no ballast without work, then what is the monetary value of that ballast? If millions of people are on welfare, what does that mean in terms of the wealth of nations? We have made a kind of Faustian bargain. Increased wealth was traditionally seen as the ability to meet society's needs both quantitatively and qualitatively; that is, more was better. The bargain was that economic growth leading to employment would have many impacts, but the perceived benefits would be so great that there was little questioning of the means required to achieve them.

But there is a new feeling abroad in the 1990s: anxiety. Just as the people in the former Soviet Union and other Warsaw Pact nations became unhappy with their existing institutions — which denied them freedom and access to the opulent lifestyles pictured on their television sets — so too did their counterparts in the West become disgruntled when economies there began to falter. Consumer confidence in the economy fell as unemployment and bankruptcies rose. Political leaders lost the trust of their constituents and business people lost the trust of their shareholders. The "uneasy eighties" turned into the "worrisome nineties."

The question is why? Did cybernetics pioneer Norbert

Wiener have it right as early as 1950, just two years after the development of the transistor? The authors of a *Scientific American* article quoted from Wiener's book *The Human Use of Human Beings*:

> The factory of the future ... will be controlled by something like a modern high speed computing machine.... Industry will be flooded with the new tools to the extent that they appear to yield immediate profits.... It is perfectly clear that this will produce an unemployment situation, in comparison with which the present recession and even the depression of the thirties will seem a pleasant joke."[10]

The *Scientific American* authors thought Wiener was wrong. But that was in 1989.

Both ethical and technological questions are critical in any discussion of the future of work. The two are so interlaced it is almost impossible to separate them. The industrial and information revolutions have radically changed the lives of people whether they live in nations of the North or South. Whatever system of evaluation we use, the magnitude of the impact has been enormous. No appreciation of the nature of wealth is possible without taking into consideration how society has been affected not only by the nature and power of the new tools, but also by the speed with which they have been introduced.

Impacts of the Industrial and Information Revolutions

If a revolution can be defined as a very large change in a very short time, then technology emerges as a major revolutionary force. Whether we like it or not, technology is being developed and implemented at an ever increasing rate.

The first part of the Industrial Revolution was characterized by the replacement of muscle power — whether human or animal—with machines. This is probably why we still measure the output of many machines in terms of horsepower. The Industrial Revolution happened almost accidentally. Scientists discovered that water converted into steam, and water power converted to electricity, which led to increases in output. These new forms of power meant that production could be located closer to markets. New types of products began to appear that were dependent on the new industrial technologies. The invention of the internal combustion engine helped to speed developments. The European wars and the resulting demand for new weapons, equipment, and communications systems also helped spur the introduction of new technologies. New ways of solving old problems could be found with the application of technology, and existing technologies (many based on past scientific discoveries) were soon exhausted.

The development of technology was just one of society's many activities one hundred years ago. Today the constant improvement of technology is the central activity of society. Especially since the end of the Second World War, the role of research and development has been seen as central to the health and well-being of industrialized nations. The pace of discovery and application is increasing. The diffusion rate of new technologies is increasing. The lag time between discovery and application is decreasing. For example, consider the electronic switch, which has become vital in telecommunications. By 1978, a mere fifteen years after its introduction, only a handful of countries had not yet been able to adopt this electronic innovation. This is a rapid diffusion rate, especially when compared to the international dissemination of earlier major technologies such as continuous casting, basic oxygen steel, float glass, shuttleless looms, and open-ended spinning.[11]

There is an international race among the industrialized nations to discover, produce, and implement new technologies.

Whereas technology was once developed by individuals in an unselfconscious way, today the fortunes of nations rise and fall based on the output of teams of researchers. The motto is "fast is good, faster is better." International bodies such as the OECD (Organization for Economic Co-operation and Development) turn out report after report trumpeting the benefits of more innovation leading to more technology, leading to more productivity, leading to ever greater economic growth. A 1980 report, *Technical Change and Economic Policy*, contained a typical OECD recommendation: "Technical advance cannot be taken for granted. Neither the rate nor the direction can now be regarded as satisfactory. The rate has slowed down substantially. . . . It is lacking in some areas where it is vitally needed." Similar exhortations can be found in policy documents emerging from conventional economic policy institutions in any year from 1970 to 1990.

Most countries have adopted costly policies to spur the diffusion of new technologies. Between 1984 and 1989 Denmark spent U.S.$145 million to encourage the diffusion of information technology in its small industrial sector. As long ago as 1979 the United Kingdom launched a Micro-electronics Application Project to encourage the use of information technologies in industry. An initial budget of £55 million was increased to £85 million by 1982. France, Sweden, Germany, and Japan launched similar programs. The main thrusts are the development of product, software, and skills. Virtually all OECD countries have technology diffusion programs of one sort or another in place.[12] Even though the impacts of industrial development are little studied or understood, the race to apply still more technology to achieve still more output continues apace. So obsessed with technological progress are the nations of the world that one researcher has referred to the trend as a type of secular religious activity.[13]

The impacts of technological change, taken together, have

shaped our current society. One example is the specialization of labour to achieve higher production. From Adam Smith to Henry Ford there has been a growth in output as productivity levels have increased with specialization. But this has also brought with it a narrowing of view. Workers, participating in only part of the production process and unable to see the final product, quite correctly came to see themselves as cogs in an industrial machine, and values such as pride of workmanship quickly gave way to a sense of alienation. This aspect of industrial production was epitomized by the assembly line and graphically illustrated in the Charlie Chaplin film *Modern Times*. The information era continues to value specialization of function. Most companies and government offices have teams of people entering data with factory-like efficiency. Teams of people write software. In most cases the individual is responsible for only a small part of the program.

The rise of specialization led to increased economies of scale, and the price paid was a decline in job satisfaction. Because workers never created a finished product, they could never see the value of their efforts. Lacking job satisfaction, they had to be satisfied with their weekly pay packet. Because the only reason to work was to exchange labour for money, the next logical step was the formation of the trade union movement. If work was only about money, then why not organize collectively to secure the highest possible hourly wage?

As individual workers became a small part of a larger production process, they also became interchangeable. No longer was it an individual with a farm or craft who produced a product or brought something to market. Now it was a matter of any number of "faceless" individuals who could produce the amount of labour inputs required by the production process. The emergence of terms such as "man-hours" or "man-years" reflected the view that labour had become a factor of production, with units of labour bought by the hour or year. The actual identity of the individuals was of no consequence. Like horsepower, the amount of man-hours expended in producing a product became a measure of worth, or cost, or value.

The growth of an industrial society took labour into the factory/office, and the creation of a mobile working class led to a separation of the workplace from the home and the home community. People left their homes in the morning, only to return in the evening. Energies and interest that might have been focused on maintaining or bettering the local community became scattered.

The information revolution could herald a return to home-centred work. The value of "telecommuting" or of the "electronic cottage" is obvious. Savings in commuting time, less energy used in travel, and less expenditures on highways and transportation infrastructure are some of the benefits. A possible problem is a blurring of work and home. Rather than humanizing work we may find that the values and structures of the office/factory migrate to the home setting. Paradoxically, rather than humanizing the workplace we may end up dehumanizing the home.

A parallel occurrence is the emergence of the family as an economic unit. In the pre-industrial situation one or more family members interacted with the money economy by bartering or selling goods and services. Tasks at home were carried out by aunts, uncles, children, and grandparents. The rise of the industrial marketplace led to the interaction of most if not all members of the household with the money economy. All members of the family who can do so sell their labour skills for money, which is then used to buy goods and services in the larger marketplace. But the tasks that used to be carried out by family members must now be paid for. The multiple wage-earning family has ended up paying for child care. Many also pay for house-cleaning services or for caregivers to attend to the needs of older members. In short, we have witnessed the increasing "monetization" of the family. Even the choice to have children becomes an economic decision, because children affect the family's income stream. One of the wage-earners may have to at least temporarily drop out of the labour force. Food costs increase, and money will be needed to clothe and educate the child and perhaps provide for increased housing space.

The spread of the money economy has also meant that to purchase more goods and services people had to earn more. Over time we've come to equate material accumulation with increased well-being, and we've had to run faster and faster to keep up. If self-worth, for individuals as well as nations, is equated with having more this year than last year and/or more than one's neighbour, it is clear that the economic race will only lead to the exhaustion of people and nations.

As people increasingly came to depend on outside institutions where they could exchange their skills for money, a feeling of vulnerability arose, accompanied by a fear of loss of control. In an earlier agrarian economy, the control point was seen to be of a higher order and appeals to religion or natural spirits could be undertaken to ensure good weather for crops or a good hunting season. In some industrial situations the locus of control was easy to identify: it was the boss, the bank, the mortgage company, or other nations taking our jobs or invading our markets. Some developments seemed to hit both workers and owners alike. Business cycles, new technologies, and imports could lead to downturns, plant closings, and job loss. Again the role of the trade union movement is easy to understand in this mix of events. Unions bargained for job security, ever higher wages, an orderly introduction of new technology, and safe working conditions.

Because more is seen to be better, it is only logical that a government should see its major role as tinkering with the economy to make it grow faster. Elections are fought and won or lost on who promises to bring about the fastest rate of growth, and Western societies have indeed achieved impressive rates of growth. With little regard for the future or the environment, we have created societies that are awash in goods. Because the environment was seen to be something that could be exploited for its resources and used as a dumping ground for the unwanted byproducts of the production process, we have come to a time when we are goods-rich but environmentally poor. The transition to an information society seems to promise, at least, less stress on the environment.

As our society industrialized, the number of people in the agricultural sector declined as people moved off the land to work in factories. In highly industrialized countries such as the United States and Canada, only about 5 per cent of the population directly produces food. The industrialization of the agricultural process led to a decline in the need for people to work on farms. But with the tremendous demand for automobiles, stoves, refrigerators, radios, and a host of other products, employment in factories increased dramatically, and the transition was made to an industrial society. In turn innovations in the manufacturing sector led to a displacement of labour, and the service sector began to grow to its current level of just over 70 per cent of the working population, with only 20 per cent left in manufacturing. With such a high percentage of the workforce in the service sector, it is said that we are now living in a post-industrial society or the information age.

One early characteristic of the service sector was the relatively high degree of labour intensity. A bank clerk or a postal worker, a secretary, an architect, or a draughtsman could only work at a certain rate of production. The automation of the sector involves the substitution of machine intelligence for human intelligence, and as a result we are now, have been for some time, and will continue to be substituting machines for work that formerly depended on human intelligence. The service sector is becoming ever more capital intensive. The capital that is being put in place is largely made up of those elements that comprise the information infrastructure.[14]

We are moving from a tangible to an intangible economy. One effect of the information age appears to be a move away from things to experiences. Playing video games, watching video movies or television, and computer conferencing are activities that seem to be on the leading edge of such an economy. That people will pay to experience implied reality means that an economy based on intangibles is being put in place. We are witnessing the emergence of content as a commodity or, put another way, we are seeing the "commodification of content." Whether this is good or bad is left for the individual to consider and decide.

Software and Network Intelligence: The Key to an Automated Service Sector

In any of the practical arts it is their physical manifestations that first seem to catch the imagination. For example, in communications we tend to think of the telephone rather than the message. We think of television as a system made up of studios with cameras interconnected by transmission towers or satellites to receivers that relay programs that have been bought or made. Just as wealth tends to be defined as material things, so too is technology perceived in terms of the components that are the most tangible or visible.

The major initial costs of any enterprise also seem to relate to equipment. Conceptual elements are much more difficult to measure, because many of them are the products of the ideas of different individuals. They may have been germinating for years and only realized after a serendipitous confluence of events has made for a right-time-right-place circumstance. This is particularly true of the complex technological innovations responsible for the information revolution.

The question of the relative importance of things and thought is even more complicated when we talk about software and network intelligence. In contrast to hardware, which refers to equipment such as integrated circuits, chips, input devices, microprocessors, computers, and modems, software refers to the sets of commands and instructions that tell the computer what to do. The cost factor has played its traditional role here as well. In 1950 hardware represented 90 per cent of data processing costs, while in 1985 it was only 20 per cent and projected to fall another 10 per cent by the end of the decade.

While progress in storage capacity, miniaturization, and speed of operations has dramatically exceeded all expectations, with the costs of computation being halved about every three years, the applications of the new information technology have multiplied beyond anything conceivable ten years ago. It is the software phenomenon that threatens to make Wiener's prophecy—that the new tools will be responsible for massive unemployment—come true.

None of this is meant to downgrade the importance of the enormous advances made in the development of supercomputers and transmission technologies. The ability to store large amounts of data and to process large amounts of complex calculations in unbelievably short spans of time has made it possible to design new and better industrial products. Modern and more efficient aircraft and cars are the result not only of the software used in their design but also of the power of the computer to handle the questions and instructions involved.

Hardware and software innovations are so interwoven that it is often difficult to decide which is the cart and which is the horse. The very existence of the computers of the late 1970s and 1980s, together with digital transmissions over copper wire and optical fibres, undoubtedly stimulated the programmers and resulted in improved word processors, spreadsheets, number crunching, and various aids to design and manufacturing. But by the beginning of the 1990s the creators of software had become the driving force of innovation. They needed faster means of computation, greater band-width to accommodate the many levels of detailed information involved, and increased flexibility in communication systems to permit computers to talk to and understand one another.

As a result, scientists began to look beyond the sequential systems of operation that had been standard since the time of the first electronic computer in the late 1940s to examine the possibilities of parallel processes using multiprocessor computers. Proponents of the viability of artificial intelligence studied human cognition and concluded that neural networks had a variety of inputs and were able to process a multitude of signals simultaneously.

Paul Wallich commented on the development of artificial neurons as essential components of neural networks. He described how attempts were being made to copy the composition and performance of nerve cells in silicon with the object of creating computers that could be able to imitate human thought.[15] It was only logical, then, that attempts would be made to apply the human model to a machine.

One technique involved breaking problems into their parts and attempting to solve each part simultaneously. To achieve this, multiple subcomputers would be combined to enable parallel software systems to operate using specially developed computer languages. There were a number of successes. A machine built by Daniel Hills incorporated 65,000 processors into a computer he called The Connection Machine.[16] A group at Yale University developed a method of building large programs out of many simultaneously active ones, which enabled them to put together large software machines and the basic elements required to meet their needs. The immediate objective was to meet the requirements for additional computing power. In the longer term the hope was that this additional power could be used for new kinds of computing that could "transform mere facts into knowledge on a vast scale."[17]

Hardware and software developments continued at a rapid pace, the aim being to broaden and deepen the range of computer solutions to human problems. The building of software was and still is a very labour-intensive activity. Software writers still tend to go line by line, writing software code in the appropriate language. Later the software has to be tested and any errors removed, that is, it has to be "debugged." Over the years there have been attempts to speed up the creation of software. Early sophisticated computer languages such as FORTRAN and COBOL are still used with a range of more powerful languages such as C or C++.

The intent now is not to instruct the computer about what to do in a step-by-step algorithm. Rather, the approach taken is to define what set of actions is needed to achieve the object of the programmer and to have the software itself generate the code. The partial automating of the process of software generation is known as CASE (computer-aided systems engineering). Included in the CASE bag of tools are other techniques designed to speed up software production, most notably the ability to reuse or recycle code used in one application for a later application. Here code is treated as building blocks that can be used and reused in similar applications. Although CASE originally referred to computer-aided *software* engineering, it is now used to denote computer-aided *systems* engineering, which underlines the merging of software and hardware. One article that made this point also stressed the need for greater integration of analysis, design, code generation, and testing packages.[18]

The larger question is whether or not it is conceivable for systems to be devised that could assist in the solution of the major problems facing society. In 1970 the Club of Rome used modelling techniques developed by Jay Forrester as the basis of its pioneering book *Limits To Growth*. But those techniques, though revolutionary then, are crude by today's standards. Furthermore, the system could consider only a limited number of factors, and despite deliberate attempts to provide for interactivity the necessary computing power was too weak to achieve the desired degree of problem-solving. It is probably still not strong enough today, even though it is several million times greater.

We have already gone some distance on the road from POTS to PANS (plain old telephone service to pretty amazing network services), but there is still much to be accomplished if we are to become able not only to move from information to knowledge but also, as Northrop Frye said, to make the value judgements that will contribute in a positive way to the quality of our lives.[19]

The age of telecommunications could be said to have begun in 1844 with Morse's demonstration of the telegraph. With this invention the speed of transporting human messages leaped from the limits of messages carried by the existing transportation system (10 to 50 miles per hour including carrier pigeons) to the speed of light (186,000 miles per second).[20] There is still some dispute over who really invented the telephone. There is little dispute, however, about the dramatic developments since that time—with more forecast for the near future. A growing list of network features and functions was added to telegraphy.

Bell received the first patent on the telephone in 1876. As the number of lines grew it soon became clear that some form of switching mechanism was needed to make connections. The problem was solved by connecting users to a central office. The office itself is often referred to as a "switch," because lines are switched open or closed with one another. The early switch was the board, where the operator literally plugged different lines into one another to meet the request of the caller. This function was taken over by cleverly designed mechanical and then electro-mechanical devices. Today's switch is really a computer that makes the connection, routes the call over the most efficient path, records billing information, and helps to trouble-shoot the network if there are problems.[21]

The convergence of computer and communications technologies is transformative in the sense that it has changed how people and physical entities are able to interact. The telephone is one of the best examples of how revolutionary the process has been. From a rotary-dial instrument — usually black — that could be found in nearly everyone's front hallway, the phone has become available in a variety of forms, a variety of shapes — and with Touch-Tone capability now standard. The external changes only hint at those that have taken place within the telephone system itself, which has come a long way from electro-mechanical switches and rotary-dial telephones. Today's networks carry voice and data in a digital format — the language of the computer. At their hearts lie a very sophisticated computer technology.

Networks increasingly use electronic switches. Each new switch is really a computer, and it is the presence of these computers that makes the system "smart." In these networks the identity of the called party can be revealed at the time of placing the call. The appropriate software makes possible a number of features: the display of the calling number when the customer's phone rings; distinctive ringing for preset numbers; blocking calls from certain numbers; selective call forwarding; call tracing by the customers; and automatic callback. With a telecommunications system based on computers, the very "smartness" of the networks raises the potential for increased productivity of all transactions.

Parallel developments in carrying capacity meant that increasingly "smart" networks could carry more traffic. The original copper wire could carry a limited amount of information. This capacity was boosted when analog voice was translated into the digital language of the computer and sent along the line as a binary packet of zeros and ones. The telephone line could be kept relatively full, because as long as the packets were correctly addressed they could be broken up and interlaced with other packets from other conversations. Soon even more capacity was added. Satellites carried increased traffic. This broadband technology was designed to carry the information of many television broadcasts, so that simultaneously transmitting a large number of telephone conversations became a simple task. The price paid was often an annoying echo or delay, but this problem was solved with the introduction of fibre optics. Not only is it possible to use fibre optics to transmit a very large number of conversations plus data and video, but this can be done without static interference. So popular is fibre that we are seeing a case of a very new technology (fibre) beginning to take away the market of a new technology (satellite).

Intelligent networks allow for a variety of new services. As well, existing services can be produced far more efficiently. For example, "smart" telecommunications networks can reference, store, or communicate information on users and traffic on the network. They can access directories and databases to offer information or complete transactions. With increased integration of computers, telephones, video, data—all delivered in a digital format—multimedia devices are already moving from pilot models to production. With intelligence in the system, network offerings can be personalized. This includes having a life-time phone number (offered by AT&T in April 1992) that travels with individual people as they move from one geographical location to another. A person merely has to punch numbers into an available telephone, thus letting the network know the location through which phone calls to the designated number should be routed.*

The use of 1-800 numbers has increased dramatically, partly because of the increased service that can be provided. For example, one tire company advertises a 1-800 number to call for service. When a motorist in need of help dials that number he or she is immediately informed about the nearest tire dealer. This can be done since the incoming 1-800 call is processed through an automatic number identification. The electronic switches put in place to route the call take the caller-number information that precedes the call and the system, using a reverse directory technique, displays the address of the calling party. The system quickly matches that information to a data base of tire dealers and provides the information to the customer. A similar service is available to the owners of Cadillac cars in Canada and the United States, with the added convenience that the nearest service centre responds directly.

American Express also employs this technology. A customer's call automatically triggers the retrieval of his/her record, so the American Express agent picking up the phone sees the customer's records displayed on a monitor. When the system was first introduced the agent would typically greet callers by name, often unnerving the callers, who couldn't understand how their names were known before they had even announced themselves. If the agent can't solve a problem with the account, the

---

*1-700, 1-800, and 1-900 numbers provide access to specific services offered by various telephone companies.

phone call with the record is transferred (via phone lines) to another agent's computer.

Smart networks allow customers to share the communications space with the seller. Sharing the space allows for transactions to be speeded up. It also allows customers to do some of the work, thereby saving a portion or all of the labour costs.

In shopping for food, for example, in earlier times customers went up to a counter in a store and asked for their groceries. Clerks found the items *behind the counter* and brought them out to the customers, who were *in front of the counter*. Using this system a clerk could wait on only one person at a time. In newer supermarkets customers share the space of the food sellers, handle the inventories, serve themselves, take groceries to the checkout counters, and leave. The ratio of clerks to customers was radically changed; labour costs were dramatically lowered.

The use of Universal Product Codes and laser scanners makes the supermarket of the 1990s even more "productive" than the supermarket of recent decades. Most stores now use laser scanners that read the price and type of product purchased. This action prepares the price totals for the consumer and at the same time sends a message to the store's computer advising it of the reduction in the inventory. In line with the shared space analogy, some supermarkets are moving to a system in which the consumers draw the products over the scanner. Here, in one gesture, consumers are preparing the price totals for their groceries and advising the stores' computers about inventory change.

Another novel use of shared space is the customer-supplier relationship. For example, a manufacturer/supplier of blue jeans, such as Levi Strauss, establishes a relationship with its customers, the stores, that gives it access to the retailers' computers. By tracking the sales of different items the manufacturer/supplier can make the decision to supply additional products automatically, depending on the instructions given to the computer. This is a variant of the "just-in-time" inventory management now so popular with manufacturers.

Banks and other businesses are allowing customers to "rove" through their mainframes, do transactions, gain information on latest interest rates and security offerings, and sign off. This sharing of space means that people will, more and more, be using personal computers or the Touch-Tone buttons of their phones to gain information, book tickets, and conduct other transactions. The sharing of computer space enhances the productivity of the service sector.

The number and scale of new services appear limited only by our imagination and market acceptance. The merging of telecommunications and computers allows *everything* to be expressed in a digital format. This is why charge cards can send information via the telephone lines to a bank's computer, all in the same digital format. We have created a kind of Esperanto of the machine world—a major force leading to the global village. Intelligent networks have emerged as a major force leading to the globalization of economic activities.

The almost magical features of advanced software and network intelligence tend to draw our attention away from the reality of the workplace. The new products and services are being introduced in and around existing jobs. Some work is made more productive while other work is displaced by the new technology. When automation occurred on the farm, displaced workers could go to the factory. When automation displaced the workers from the factory, the service sector seemed able to absorb them. Now advanced software and intelligent networks are changing the service sector. New jobs will be created, but many others will be lost. The question is: where do the service sector workers go after they lose their jobs to automation?

The Automated Service Sector

If these are not the best of times, it is partly because the disparity between the rich and poor has been increasing in the industrial-ized world and the people in between have become part of a shrinking middle class. Most new jobs are to be found in the ser-vice sector, and they tend to be polarized between a relatively small number requiring high skills and the rest concentrated in the low-paying retail and consumer-service areas.

The service sector defies easy definition. Services usually refer to intangible, non-transferable, and non-storable activities. However, there is a blurring between goods and services. In fact most goods are produced and sold as complex packages of ser-vices. Harlan Cleveland notes:

> Cray computer (the supercomputer), defines the machines it produces as being composed of a large amount of knowledge and a small amount of sand ... and (according to the French economist Albert Bressand) McDonald's hamburger consists of some wheat and meat but that is the smallest part of the costs of production. Far more important are the highly developed procedures and process for production and service. Bressand then asks should the McDonald's hamburger be considered a good or a service, and concludes that it is not an important question. In the information economy what we are dealing with instead are complex packages of goods and services (which Bressand calls "compax").[22]

The Economic Council of Canada distinguished between "dynamic services," "traditional services," and "non-market services." The *dynamic* area includes four major groups:

1) transportation, communications, and utilities (including broadcasting, telecommunications, and air, rail, and water transport);

2) finance, insurance, and real estate (including banks, credit unions, insurance companies, and investment dealers);

3) business services (including employment agencies, advertising, and architectural, legal, consulting, engineering, and scientific services); and
4) wholesale trade.

The *traditional* services include retail trade, personal services, beauty and barber shops, and travel agencies, among others. The *non-market* service sector includes education, health, social services, and public administration. Most of the automation taking place in the service sector is in the area of dynamic services.

Banking and financial services provide an interesting picture of what is now occurring in the automated world. The ubiquitous automated teller machines are a constant reminder of the magnitude of the changes taking place before our eyes. As the young computer-literate generation moves into the wage-earning class, there are more and longer line-ups in front of Green Machines, Blue Machines, and the multicoloured versions that cater to networks such as Circuit, Interac, Cirrus, and Plus. Money can be obtained virtually everywhere at the push of a few buttons without the need of traveller's cheques, passports, or a knowledge of the local language. A plastic card does the trick, along with an identification number and a bank account in good standing.

For the user, accessibility is the attraction. Available seven days a week, twenty-four hours a day, the machines are found not only in financial institutions but also in shopping malls, convenience stores, or wherever people tend to gather. Cheapness and efficiency recommend them to financial institutions, since they greatly reduce the need for human go-betweens who must be paid salaries. Instead of tellers and various supervisors, only a relatively few employees are required to stock the boxes with cash and collect deposits or other paper records. These workers are primarily at the lower levels of the income scale. The amounts of withdrawals and deposits are automatically debited or credited to the appropriate accounts along with the service charges that accrue to the institution.

The net result is a decrease in demand for workers, less need for costly negotiations over conditions of employment, and fewer causes of friction between management and labour. Fewer supervisors and accountants are required, and fewer individuals skilled in human relations are needed to deal with dissatisfied customers. The audit function is enormously simplified as human error is reduced. Computers not only add, subtract, multiply, and divide without making mistakes, but are also programmed to read, balance, and update accounts and transfer money as required.

Automation has also invaded the ranks of upper and senior levels of management. The Toronto-Dominion, for example, with assets of over $74 billion in 1993, is one of Canada's six largest chartered banks. It is considered one of the most conservative financial institutions in the country but is a leader in innovative procedures designed to serve its clients. The TD Bank's trademarked Money Monitor system provides clients with instant access to their companies' current banking information through a personal computer terminal in the clients' own offices. The system enables customers to manage cash flows more efficiently by accessing current account data in both Canadian and U.S. funds, checking outstanding loan balances, monitoring current and historical account activities, transferring funds between accounts, and obtaining money-market and foreign rates — and all of this without the need to phone a branch for account information.

Another TD Bank system, Business Window, includes the Money Monitor together with other trademarked services. These other services include: ASSETLINK, which provides access to all the information needed to manage a securities portfolio; TDPX, the Toronto-Dominion Payment Exchange, which substitutes electronic data interchange (EDI) for paper-based cheques; Facilitrade, which makes provision for the automatic preparation, transmission, and receipt of Documentary Letters of Credit anywhere in the world; Fund$IN/Fund$OUT, which enables a company to collect and disburse preauthorized recurring payments quickly through an automated payment system; and Notewriter, a software program designed to let companies manage their commercial paper electronically and issue commercial paper in any quantity or denomination. Business Window has also been enhanced by a personal computer product called RAPIDWIRE for S.P.E.E.D. (Secure Payments, Excellent Electronic Delivery). This enables customers to make a one-step dial-up on the system with features that are simple, fast, secure, and flexible enough to permit browsing without the necessity of obtaining printouts.

The features introduced by the Toronto-Dominion Bank are typical of what is happening in the financial industry of the industrialized world. Indeed Canada, with 28 per cent of payments being made electronically, lags well behind Japan and Germany, where the percentages have reached 80 and 63 per cent respectively.[23] France has become a world leader in the use of "smart" or debit cards known as EFTPOS (Electronic Funds Transfer at the Point of Sale), a technology that has yet to make serious inroads elsewhere but seems sure to do so in the near future.[24]

For the financial companies the benefits are savings: costs associated with cheque production and reconciliation are eliminated; mailing costs are either eliminated or reduced; bank service charges are either lower or not as high as they might otherwise have been; the time spent on administration is shortened; cash flow forecasting is enhanced; the time and cost of preparing and handling invoices are eliminated; and bookkeeping is simplified. The companies make distinct savings in the time previously spent by workers doing jobs that are now accomplished more efficiently and in less time by machines. A report issued by Salomon Brothers referred to software that converts documents to electronic form and estimated a 50 per cent increase in productivity given appropriately streamlined work flows.[25]

Citicorp, the U.S. bank, is planning to launch a new generation home-banking terminal designed for broad consumer use. Called the Enhanced Telephone, it is a hybrid device that looks like an overweight telephone fitted with a small screen. The Enhanced Telephone is designed to stimulate the home-banking market, which is now stalled at about a hundred thousand users in the United States. The phones will be used to pay bills, transfer funds, get loans, review accounts, and, in the future, get price quotes and trade securities. Enhanced Telephones will be leased with a small installation charge and a monthly fee. The device will have a slot for insertion of a smart card. With an Enhanced Telephone in place, customers will presumably have the ultimate in home banking. They will be able to get cash at home by transferring funds from their checking or savings accounts to their smart cards, which can then be used for making purchases.

The drawbacks of these innovations primarily relate to the reduction in the amount of human contact and the dehumanization of new business processes. But, important as these factors are, they pale by comparison with the implications for the nature of work, especially in relation to the growth of unemployment that is being experienced in virtually every industrial country. In the present trend towards a global economy with the accompanying push to become competitive, a favoured buzzword is *downsizing*, a term generally used to describe reductions in costs of operations. Most of these reductions are related to labour. This means, for financial institutions and their customers, a reduction in the number of employees such as tellers, clerks, supervisors, accountants, auditors, and managers—or at least less future demand for these roles.

As part of this trend to downsizing, many corporations are revamping their headquarters. They are moving away from a preoccupation with an edifice that is both monumental and complex to one that is flexible and smaller. This shift reflects a new strategy based on fundamental shifts in the economy and management style. Technology, the thinning of middle management, and new management philosophies are making the ideas behind the big buildings outdated. A lavish use of space tends to "embarrass" corporations. One financial analyst said, "The institutional headquarters is a dodo. If I see a company building one, I sell my stock." Since the mid-1980s, as corporations have responded to global competition and technological change by merging and consolidation, downsizing and eliminating entire levels of management, some two million middle-management positions have been permanently eliminated.[26]

If financial institutions form one of the main pillars of the service sector, transportation, communications, and utilities are the others. Telecommunications has undergone the most radical change. Digitalization, lasers, and the development of fibre optics paved the way for the marriage of telephony and computers and the introduction of a host of new services. Advanced voice, data, and image communications included facsimile machines, cordless telephones, digital multiplex systems, mobile cellular phones, international automatic credit card services, full colour video conferences, and personal communication services whereby phone numbers are assigned to people rather than locations. All of these are again not without their human costs. In Australia, Telecom has reduced its staff by eliminating seventeen thousand positions in the last decade, with about six thousand of these cuts coming in 1991-92. Furthermore, there were estimates that cutbacks representing an additional 33.3 per cent could be imposed by the mid-1990s.[27]

In the United States AT&T plans to install computerized operator services, which would allow it to close thirty-one offices and eliminate one-third of all operators. This would amount to the elimination, by 1994, of about six thousand operators, all to be replaced by "voice recognition" technology that responds to a caller's verbal prompts. Instead of talking to a human operator, callers will tell the computer the type of call they want to make. The computer prompts the caller to say "collect," "third number," "person-to-person," or "calling card." When connecting a collect call the computer tells the called party, "I have a collect call from Mr. So and So. Do you accept the call?" The computer recognizes "yes" or "no" and completes the transaction based on the response from the called party.[28]

The equivalent of automated bank tellers is coming to the car rental industry. Budget Rent A Car is experimenting with remote rental booths at such unconventional sites as shopping malls and hotels. Customers use a video booth to dial a reservation agent, then punch in their charge card and driver's licence information. The machine dispenses a car key. The rental car is in a nearby parking lot.

A growing number of equipment suppliers are diagnosing customers' problems from service centres hundreds or even thousands of miles away. Increasingly, the remote diagnosis is followed up by a remote repair job. Among the companies that rely on long-distance telephone lines for diagnosis and repair are Pitney Bowes for its FAX machines, General Electric for its body-scanning systems, and AT&T for business telephone exchanges. Remote diagnosis and repair systems allow equipment suppliers to sharply reduce the number of service calls they make. They also cut down on the time spent on problems that must still be dealt with on site, because the remote diagnosis allows repairers to be dispatched with the proper tools and parts.

Pitney Bowes receives close to thirty thousand calls a month from customers at its National Diagnostics Center in Florida, where twenty-two engineers are on duty twenty-four hours a day. Most calls involve small problems, and the engineer can talk the operator through the solution over the phone. If service is needed, the diagnostic centre can often fix the problem remotely. Through a phone line the customer's machine is ordered to send an electronic status report covering a hundred or more features. The report takes twenty seconds and is used to program a Florida FAX machine into a duplicate of the customer's. A list of the last twenty transactions is also received by phone from the customer's machine. Problems show up with error codes attached to the transaction. The repairs are made to the software in the machine in Florida, and the fixed software is sent back through the phone lines.

In the near future problems affecting appliances such as refrigerators, air-conditioning units, and washing machines will be remotely diagnosed in the home using phone lines. It is cheaper to determine the nature of the problem remotely and talk the home owner through a minor repair than to send a service person to the site. However, if necessary the expert can arrive at the home with the proper part and advance knowledge of the precise problem.

Accordingly, sometimes the labour saved by corporations is shifted to the consumer. A few years ago McDonald's came up with the slogan, "We do it all for you." In reality, at McDonald's we do it all for them. We stand in line, take the food to the table, dispose of the waste, and stack our trays. As labour costs rise and as technology develops, consumers often do more of the work. Microprocessors allow for self-serve gas stations and automatic teller machines. The laser scanners being introduced in supermarkets are going to force consumers to pass products over the scanners and thereby save companies the cost of employing cashiers. Already, one telephone company allows householders to become telephone installers. New residents in the State of Washington can obtain telephone service in a self-serve way. They plug a Touch-Tone phone into a jack and press 811 on the telephone key-pad. A computer asks questions that the customers can answer by pressing more digits. With this system a phone is operating in less than twenty minutes, compared with the usual two-day wait until a technician arrives. In Washington, when university students returned to campus in September the "service on demand" system was able to handle a surge of 1,726 orders in two days instead of the usual two weeks.

Audiotex is another technology that allows us to speak into phones and have our voices captured on voice-recognition chips. Our digitized voices can be delivered to someone else someplace else in the network. Using the Touch-Tone telephone key-pad, we punch a series of numbers to work our way through an information base to learn something, to express a view, or to purchase a product. Audiotex, or info-bots (as in robots), can answer phones, route callers, and dispense information. The more powerful units combine voice-message recognition with computer files to enable callers to use their telephones to guide them through an information data base.

Just as the new telephone technology is replacing human secretaries and telephone operators, sophisticated software is replacing humans throughout the corporate world at all levels of the hierarchy. For example, expert systems or decision-support software are now part of the everyday business world. Shearson Lehman uses neural networks to predict the performance of stocks and bonds. Merced County in California has an expert system that decides if applicants should receive welfare benefits. A U.S. telephone company has a system that helps unskilled workers diagnose customer phone problems. The U.S. Internal Revenue Service is testing software designed to read tax returns and detect fraud. American Airlines has an expert system that schedules the routine maintenance of its airplanes.[29]

The computer applications represent the knowledge of humans that has been transferred or embedded in the software. The opportunities are immense, because most procedures and processes depend on an ever increasing level of expertise. One estimate is that the applied research related to how manufactured goods are built and how they work makes up 70 per cent of their development costs. In the service sector, in areas such as the selling of mutual funds, the percentage rises to 90 per cent. The incentive to transfer knowledge from human brains to computer software is very high indeed.

Automating the service sector produces faster service and makes labour redundant. For example, the Merced County system is able to review a matrix of six thousand government regulations to determine in seventy-two hours — versus as long as three months — if an applicant qualifies for benefits. Because the intelligence is in the machine, the system requires a smaller support staff and less skilled interviewers. The system is reputed to save the county $4 million per year in administrative and training costs. The county welfare agency has also been able to cut its staff by 28 per cent and still serve the same client caseload.

All of these innovative services could not have been achieved without the corresponding advances in telecommunications. Nor could they have been offered without the new and powerful software packages. Digital technology plus modern telecommunications has led to a revolution in the way services are delivered. The new technologies based on microelectronics are all-pervasive, which serves to remind us that we live in an interdependent world in which every change has a multiplicity of impacts. Every new service offered means savings of one kind or another for either the supplier or the customer. Businesses that contract for a cash management or sophisticated electronic data system benefit in a material sense. Whether or not they employ tellers or operators as such, they do require the work done by accountants, auditors, and record keepers, and it is these positions that they are able to eliminate. The instantaneous access to information about accounts receivable and payable reduces the need for large finance departments with a staff of middle managers. The downsizing also means that there is less demand for office space and hence less demand for construction workers, carpenters, plumbers, painters, designers, and architects. With fewer available jobs and reduced payrolls, does this mean that there are fewer people with the income required to buy the products of the now more efficient and competitive retailers and manufacturers?

The automation of the service sector is moving quickly — personal computers, e-mail, local area networks, FAX machines, new software applications, pages, cellular phones — the list goes on and on. The net effect is always the same. New "efficient and effective" services are offered, and the remaining workers supposedly become much more productive. So why worry? Isn't it nice to have our tastes and interests profile lodged in a computer data base somewhere so that when we phone a hotel for reservations the computer "remembers" that we want a particular type of room, located in a particular section of the hotel? Or that when we rent a car the agent can have access to our previous rental file and get us on the road in minutes? The worry is that the automation of the service sector, while providing a dizzying array of new means of satisfying our wants, is also having an impact on the kinds of jobs that are available.

We are beginning to see the creation of a workforce characterized by a bi-modal set of skills. Highly trained people design and implement the technology, and unskilled people are hired for the remaining jobs. There appears to be a widening skills gulf. Even more disturbing is that income patterns reflect this divide. The middle class is shrinking, with only a few moving up to the upper class and many more taking the step down.

Added to the skill/income problem is a parallel rise in non-standard employment. This includes part-time work, short-term work, and self-employment. Most part-time workers are classed as "involuntary" part-time, which means they would prefer regular jobs. In Canada non-standard jobs account for nearly half of all new jobs and now represent nearly 30 per cent of total employment. Non-standard jobs tend to be non-union, less likely to be covered by employee benefits, and without pension plans or other benefits.[30]

While not directly caused by information technology, much non-standard employment is related to the trend to "contract out" work. With information technology certain kinds of work can be done anywhere and purchased as needed by other firms. The influence of labour unions is minimized, overhead costs are kept down, and the labour, when acquired, is cheaper because there are few, if any, employee benefits to be paid for.

The overall effect of automating the service sector is still unclear. What is known is that for a variety of reasons, including automation, there is a polarization of incomes taking place. In 1967, 27 per cent of the Canadian workforce had annual earnings that were middle level (within 25 per cent of the median on either side). By 1986 only 22 per cent of the labour force fell within this group—a decline of 5 per cent. The low-income segment rose from 36 per cent in 1967 to 40 per cent in 1986. The high-income segment went from 37 per cent in 1967 to 39 per cent in 1986.[31]

There appears to be clear evidence of a "declining middle." While the experts differ on why the middle class is in decline, there is general agreement that a continuation of this trend threatens the social fabric of Canadian society.

Similar trends appear to be under way in the United States as well. Between 1970 and 1990 the share of all income received by the richest one-fifth of families grew by 3.3 percentage points to 43.7 per cent. The share received by the poorest one-fifth fell by 1 percentage point to 4.6 per cent. The income shares of families in the second and third quintile also fell, and the share received by the fourth quintile—the fifth of families just beneath the richest—rose slightly.[32]

When big corporations began laying off white-collar employees in the early 1980s, 90 per cent were quickly re-employed in similar jobs in large companies at the same pay or better. In the late 1980s only 50 per cent were rehired. In the 1990s only 25 per cent are able to come back into the corporate world. According to *Business Week*, quoting one executive hiring firm, "There just aren't those jobs any more, and they can't hope to ever get them again. . . . They're going to have to think of other ways of being employed."[33] While middle managers represent only 6 to 7 per cent of the United States workforce, almost 17 per cent of corporate layoffs between 1988 and 1991 came from their ranks.

Our society is going through a wrenching change. From a time when one wage-earner could support a spouse, clothe, feed, and educate two or more children, and buy a new car and home, we seem to be moving to a two-tier society in which the poor and the rich will see each other across a great unbridgeable divide.

The Boeing 747 aircraft might provide an appropriate analogy. In one part of the multimillion-dollar machine are the highly skilled pilot, co-pilot, and navigator. In the other part are the stewards and stewardesses. With this bi-model distribution of skills one cannot become a pilot by working as a steward or stewardess for the same airline. Rather, one has to drop out and, if possible, be totally retrained for the more skilled and highly paid job. The skills profile of our society will tend more and more to resemble this analogy. One group of people will push buttons, monitor lights, and make simple repairs. Another group will design the hardware and software that make the system function. In between there will be a declining demand for the semi-skilled worker. Broadening the 747 analogy: what about the passengers who are transported, fed, and amused? Is this the most likely outcome for the great mass of our population who are no longer able to find satisfying work or any kind of work at all? Will we have to provide games, movies, food, alcohol, and drugs to the unemployed mass of the population to keep them amused through the journey of life?

With the automation of the service sector and the decreasing amount of available work, a major problem arises in the notion of work itself. What is the function of work? Originally work was needed to grow food, make products, and build houses, cities, railways, and highways. People worked or were paid to work because they participated in the transformation of a resource into a commodity or sold a personal service in the marketplace. The process of work and employment was a means to an end. Over time work has changed from a needed activity that transformed resources into usable goods to a mechanism for distribution of income. Self-esteem, identity, and self-worth have meshed with the employment of the individual. People still tend to describe themselves in terms of what they do for a living and not in terms of their hobbies and interests. How will our tendency to define people in terms of their work and what they *do* change those who no longer *do* anything? How will people behave in a largely automated world? Today the means has become an end, and the political mechanism has taken upon itself the task of creating jobs as a method of distributing income. With many jobs becoming redundant we must find new and creative means for distributing income.

We live in industrialized societies that seem to have achieved many of their basic objectives, particularly the general provision of goods and services and a high standard of living for much of the population. We still wrestle with questions such as the meaning of it all, but no economic system can provide an answer to that mystery. One irony of the success of capitalism is that, in some respects, the capitalist system has reached the theoretical end point of communism — that is, a situation in which many goods and services are potentially "free." However, while mature and successful capitalism has apparently solved the production problem, it does not have the ideology or institutions necessary to enable the full distribution of automated goods and services to all people. Communism had the ideological mechanisms for distributing goods and services, but couldn't solve the production problem.

Our societies seem to be locked into a system of attempting to create more jobs so that money can be earned to enable the people to buy the goods. As we have seen, however, with automation even job creation is becoming less and less possible. With fewer people working, tax revenues are declining and there is a decreased distribution potential. We are facing a crisis. If we allow the system to go on in a "business as usual" way we face a set of problems that will, at a minimum, radically alter our sense of well-being. The probability is that we are inviting an economic collapse. What are we to do? How are we to devise policies for dealing with the potential abundance of a largely automated structure? What is the policy role for governments in an information economy? And if we can devise the policies, are we mature enough to implement them?

Implications for the Future: Some Scenarios

Technologies driven by microelectronics and concomitant developments in communications have promised a new era that would be highly productive and environmentally benign, and would make material goods accessible at prices so low that virtually all members of society could enjoy a quality of life never before possible. What is more, work was to be transformed from an activity both tedious and backbreaking to a rewarding experience that would meet not only the primary needs for food, clothing, and shelter but also those secondary though essential human requirements for variety, recognition, and an experience of success.

But the microprocessor, thought to be "the engine of the eighties," appears to have sputtered in the nineties. As a writer noted in *Fortune* magazine, "No novelist would dare put into a book the most extreme of the dizzying contrasts of wealth and poverty that make up the ordinary texture of life in today's American cities."[34] In Third World countries, millions have fled their homes, infant mortality rates are increasing, and poverty is at an all-time high. The industrial nations are experiencing the worst economic downturn since the Great Depression of the 1930s. Unemployment is rampant. Social unrest is increasing, and many young people have given up hope of enjoying a standard of living equal to that of their parents or even their grandparents. Furthermore, as a United Nations report said, "Environmental destruction is increasing."[35]

Just as Shakespeare's Hamlet declaimed, "To be or not to be: that is the question," we might well ask, "To work or not to work?" What has gone wrong? The analogy with *Hamlet* is not inappropriate, because indecision is now the predominant attribute of many of our political leaders. Traditional policies seem inadequate in light of the magnitude of the problems faced.

The nature of work has changed as a result of the introduction of new tools and, in particular, of those associated with the Information Revolution. Much of this change has taken place as a result of the automation of the service sector, including the jobs eliminated when new forms of hardware and software were introduced. In the past new employment opportunities in the service sector largely made up for work lost following the automation of manufacturing. Now the sector of last resort seems to be welfare, which a declining middle class appears increasingly unable or unwilling to support through higher taxes. If Stendhal was right when he referred to work as the essential ballast in the vessel of life, reliance on welfare could well lead to the destruction of life as we know it.

But the problem is not just technological in nature. The issue is complex: partly economic, partly ideological, partly political, partly sociological, partly psychological, and partly just bad luck. The poet Alexander Pope said, "Know then thyself" — not bad advice when applied two and a half centuries later to the question of what is now going wrong. Debt, both at the national and individual levels, is surely a contributing factor. Collectively and individually we tend to spend more than we can afford to. Politicians in the major industrial nations say their hands are tied because such a large portion of government revenues must go to pay the interest charges on government debt. Corporations are similarly overextended. Individuals have been seduced both by advertisers and by the instant availability of credit through the use of the plastic cards they have been encouraged to acquire from banks, trust companies, and retailers. Accordingly we tend to set little or nothing aside for the proverbial rainy day.

With the collapse of the Soviet Union the advocates of the so-called free market have had a field day. Their rallying cry has been deregulation and the need to compete in a global economy. In a version of social Darwinism they have preached that the survival of the corporate fittest promises cheaper prices to consumers, and that this will result in new economic demand, new industries, new jobs, and new wealth. The fallacy in this ideological argument is the failure to take into account the lack of purchasing power of the dispossessed during a long and painful transition period.

Politicians in democratic states have had particular difficulty in coping with the rapidity of the changes taking place as a result of the Information Revolution. Their time in power between elections is usually not more than four or five years, which inevitably means that short-term needs take precedence over long-term requirements. There is no time to consider strategies that could take advantage of the rapidly emerging technologies and the increase in the aggregate amount of goods and services available. Band-Aid solutions are sought, and surprise is the result when those solutions do not seem to work.

The changing status of women, together with the transformation of the traditional family in the Western democracies, has placed additional stress on men, women, and children as they strive to contend with single-parent relationships, child-care problems, financially strapped educational institutions, and hospitals unable to meet demands that formerly would have been cared for in the home. Family planning and medical breakthroughs in successfully treating formerly fatal diseases have resulted in an aging population, which adds to pension and health-care costs. One significant result has been a major shift in values towards a more materialistic philosophy and a drift away from those preached by humanists or the churches and other religious bodies.

These sociological changes have had serious psychological repercussions. Suicide rates among the young have climbed. More and more adults are being treated for depression and anxiety. Resentment and disenchantment with a social system that appears to be unjust are being channelled into increased use of drugs, violent behaviour, and a degree of despair hitherto unknown. Lack of confidence in and respect for political leaders has risen to such an extent that it is becoming increasingly difficult to attract highly qualified individuals to run for office in what used to be one of the most respected of professions.

The media are also partly to blame, even though it is not the tool itself but the use made of it that is inherently harmful. By and large television, with a multiplicity of channels and programs available, caters to the taste for the trivial. The ubiquitous screens in our homes depict more and more sex and violence. More often than not news and public affairs programs feature headlines rather than analysis. The concentration of ownership leads to single rather than multiple points of view. The increasing lack of willingness on the part of governments to adequately fund public broadcasting leaves the field open to advertiser-supported networks that are frequently constrained by the opinions of their clients.

And then there is just plain bad luck. Who could have foretold the plague that has resulted from the AIDS virus? Who could have foretold the increase in natural disasters such as floods and earthquakes? Who could have foretold the damage done to the environment by carbon dioxide emissions, chlorofluorocarbons, and other chemicals? Certainly not the inventors of the cars, the planes, the air-conditioners. Certainly we should have reacted sooner once the effects were known, but it takes time to know and understand what those effects are and their implications for society as a whole. Time in the age of information seems to be at a premium.

The issue that continues to confront us in this era of the Information Society is how society's wealth can be distributed to avoid collapse and, at the same time, how to avoid the errors made by those idealists attracted by the concept of the classless society expressed in the works of Marx and Engels.

If Illich was right in his contention that the tools we use must provide us with the ability to express ourselves as well as enriching the environment with the fruits of our vision and labour, then we must be able to control those tools. Without this control, how can people have a sense of meaning, of purpose, of identity and self-worth when automation becomes more and more complete? As automation increases we see an enormous scramble for the remaining jobs. To counter this scramble there is a call for more training and skills development. We need these skills, but we also need something else. We need knowledge and the understanding that is inherent in the concept of the liberal arts. We need the skills that will help people cope with a life in which they will, more and more, have to invent their own identities and discover what is worth doing.

If the situation is serious, then what paths and policies can we follow that would enable us to cope with what one editorial writer termed "The recession that won't go away"?[36] That writer offered a number of explanations for why people believe the recession will linger on, including the restructuring connected with the introduction of the new technologies, but, understandably enough, he did not offer any solutions. Undaunted by this, we have selected three possible scenarios, out of any number that could have been examined. There may be many others as equally plausible and perhaps more persuasive, and any discussion amongst informed people will inevitably bring these out. We offer the following with the full realization that they are far from being the last word.

The Business-as-Usual Scenario

"Business as usual" is a phrase common in North America. In a sense it refers to the status quo or the existing state of affairs, but whatever expression is used the sense is to carry on with existing policies and procedures. During the Second World War the British were said to have "muddled through" a variety of challenges, both economic and military. The fact that they were able to do so without inevitable destruction gave the phrase a certain cachet. The first reaction, then, to the difficulties faced in adjusting to the repercussions resulting from the replacement of people in the workforce is to do nothing or, at the most, to try and ameliorate the most serious consequences by treating the symptoms.

Under the present circumstances automation, whether in the goods or services sectors, leads to more and more unemployment. As consumption decreases, inventories increase. Consumer confidence falls along with the purchasing power of larger and larger segments of the working classes. The economy experiences a continuing recession or depression. A shrinking number of people work in situations with lifetime employment that provide them with benefits and pensions. Part-time employment and other non-standard employment are the lot of an increasing number of people in the workforce. Corporate offices are staffed by a small central co-ordinating group, with everyone else on a personal contract or having some form of an indirect relationship, such as through an intermediary prepared to provide a service on demand. Like the metaphor proposed by prominent management theorist Peter Drucker, the institution of the future will look more like a symphony orchestra or a hospital, with the central office the conductor and the players a group of contract employees.

As more individuals fall into the social security net designed to help them contend with what is presumed to be a relatively short transitionary period, the costs of the services rise and debt-ridden governments scramble to cope with the ever increasing demands for funding. Adding to the problem is the need to provide financial assistance to overextended corporations, which, despite their commitment to a system designed to reward "winners," are reluctant to become "losers." The dilemma facing governments of all political stripes is that if they don't help the corporations, massive numbers of jobs will be lost, which only increases the severity of the economic downturn. In campaigning for election in 1988 President George Bush of the United States, when promising not to impose new or additional taxes, said, "Read my lips." Despite what the "lips" were purported to have said, increased taxes, together with reductions in services, seem to be the only alternative left.

These are far from attractive options. The automation of the service sector has resulted in a shrinking middle class, and the members of this group have provided the largest share of the tax revenues in the past. Larger levies on corporations are possible, but in a global marketplace companies can take their business or declare their profits elsewhere. And many do, moving not only head offices but also factories that can take advantage of wage rates that are minuscule in comparison with those in the industrial world. The largest firms are international, and their allegiance is not to the nations that spawned them but to shareholders who may live in any one of a number of countries anywhere in the global village. There is also a management agenda to be met. Increased earnings make the stock of the company more attractive, leading to future stock offerings. Stock option purchase packages, which are becoming more and more a part of the compensation of senior management, become more valuable, thereby enriching professional executives. Where the earnings are made is irrelevant.

As the cry for globalization increases, few people question whether it is needed or who benefits and who pays. Economic theory postulates that with globalization per capita incomes will rise throughout the world. But the same theory says very little about what happens in any particular part of the world. Harmonization of wages is one outcome, with the more developed countries meeting the less developed somewhere in between. The environmental standards of the less developed areas allow for low cost production, so here again there is pressure on the industrialized world to lower environmental protection standards. Established social services such as health care, education, and libraries are allowed to deteriorate or are placed on a "user pay" basis. The principle of universal availability of services, one of the hallmarks of advanced societies, seems untenable. The rich do very well in a world of private security systems, vacation resorts, and tax breaks, but the poor are pretty much on their own, moving from one de-skilled job to another with little chance of breaking out of the poverty cycle. With the middle class disappearing as a way-station to advancement, people have few incentives to even attempt to better themselves. The elite, clinging to outmoded ideologies, are either unwilling or unable to change. As the situation worsens, reaction becomes stronger and denial the predominant mood.

The business-as-usual scenario may well be a likely one, but it is far from promising. With little or no job security we are essentially re-creating a new feudal system in which the landowners have been replaced by Kafkaesque authorities and a senior managerial class with close ties to the government of the day. The prospects for full democracy and equitable living conditions all look increasingly dim. As frustrations increase, as the dream becomes a nightmare, rioting is in danger of becoming a normal phenomenon. "Law and Order" becomes the politician's catchphrase, as the guiding principle thought best able to appeal to a shrinking number of constituents, at least until a new deliverer appears on the scene. He or she may not be a Hitler or an Idi

Amin, but will carry the alluring promise of security and full employment. All that will be asked of us in exchange is our freedom.

The Resurgence of Humanism Scenario

"The resurgence of humanism" is a positive scenario, but in some ways it is too positive. It is utopian. It involves a revival of humanism but with "new age" values and a willingness to confront moral issues.

With this revival comes a shift away from viewing society and the economy as a "win-lose" game. More for one person used to mean less for the other. When people view society as a community, they tend to perform acts and deeds without regard to payback. To some degree they shift away from materialism. The change could be associated with religious values (reward in the afterlife) or ecological values (preservation of the planet) or a new age philosophy with an emphasis on helping others, or a combination of all three. The adoption of ecological values moves away from the secular religion of economic growth at whatever cost, itself becoming a new secular religious activity.

Central to this scenario is a concern with the social impacts of technology (who pays, who benefits) — a move away from the need to adopt technology at the most rapid rate possible unless the benefits can be clearly identified — and a growing wonderment as to why society has been so preoccupied with the new and the novel at so great a cost to so many.

Allied with a growing emphasis on the community comes a tendency to move away from globalization. Even though global competition tends to lower the prices of goods and services, the humanism scenario emphasizes protectionism as community values take precedence over consumerism. The analogy is the family. Decisions affecting the family are usually taken with regard to the needs of the family as a whole. More often than not a decision taken can have a negative economic impact, but it still helps to hold the unit together. For example, a family member turns down a higher paying job in a geographically distant area because the move would go against the larger interests of the other family members. Or someone might keep rather than sell a heirloom because it has value in and of itself and helps to define the family. In short, there is a move away from the market model as the predominant paradigm.

As a result there is a rise in learning for its own sake rather than as a deliberate career-building activity, a rise in pure research and a deeper concern with culture and the arts. In effect, we witness the move away from a society in which the bottom line is measured in purely material terms.

Technology remains important but its assessment focuses on labour displacement, quality of work, and medical hazards, and appraises the overall effect of new technology on the integrity and long-term viability of the community. Paramount is the concern for long-term effects, specifically the physical and psychological health of the community and the ecological future of the planet. The approach emphasizes policies that provide for rewarding work, shorter work weeks, longer vacations with paid leave for community service, and retraining as necessary. The scenario encourages spiritual needs. The voluntary sector comes into its own. Helping others becomes a fully accepted way of life, and people don't continually ask, "What's in it for me?"

Technology recedes as a threat because it is introduced more slowly with a view to the full impacts of its implementation. The countries that appear to gain an advantage by maintaining a "technology at whatever cost" development strategy are viewed with the disdain and disapproval of a new and effective United Nations. The sanctions may be moral rather than economic, but are none the less effective. When it is determined that they have damaged the local and global environments they are called to account.

In this scenario technological progress and economic growth are slower. The emphasis shifts to the distribution of wealth and to how more services can be made universally available. It is understood that equality of access tends to minimize social frictions. The focus is always on determining what is the best technology for the community in terms of viability, social harmony, and the avoidance of sharp differences in the ability to gain access to the world's goods. The result is the creation of many more public goods that are available to all.

Elements of the humanist scenario are already in place here and there around the globe, particularly in less developed regions. But while we might applaud the scenario, it is reminiscent of Thomas More's imaginary island Utopia, so ideally perfect — and so drastic a turnaround — that it seems unlikely to be adopted in the near term.

The Enlightened Self-Interest Scenario

A less idealistic scenario is more likely to take hold. It relies on the principle of "enlightened self-interest." It is pragmatic because it is based on a recognition that the full costs of development are, sooner or later, borne by all. In this case all members of society realize that they have a stake in the benefits and costs associated with the introduction of "the new tools." They realize that, while *they* may not lose their jobs when a new technology is introduced, the emergence of a huge underclass with little or no purchasing power not only threatens the economy, but also personally threatens themselves, their families, and their communities.

In this scenario the automation continues at a high pace, but there is a continuing focus on what happens to the workers displaced by machines. The workers are seen to be part of the community; they are counselled, retrained if necessary, and not made to feel like losers.

Economists come to understand that the productivity gains inherent in the new technology must be distributed to the community to maintain an effective demand for goods and services. It is recognized that before the workers were displaced they paid taxes on their incomes. With displacement the productivity of the workers is now to be found in the productivity of the technology. Using the enlightened self-interest approach there is agreement that *some of the gains* in productivity must be taxed to create and maintain the infrastructure and universal services such as health care, and provide a basic level of income.

In essence the new levy is a kind of "technology productivity tax." But it is not one that is easy to accept. Owners of the technology/capital complain that we are killing the golden goose. The answer is that it is imperative that advances in productivity be distributed more broadly if we are to avoid economic collapse. In the end enlightened self-interest takes over, and agreement is reached. Much like collective bargaining today, owners of technology-intensive capital sit down with governments to determine tax rates. The adoption of such a tax temporarily affects the first country to legislate it, but other countries quickly see it as both necessary and desirable. The industrial nations agree, and slowly but surely the international agencies do so as well.

The technology productivity tax provides the basis for new forms of employment designed to enhance communities, revitalize aging infrastructure, and provide new infrastructure where needed. The enlightened self-interest approach also accepts that, while people are being replaced by machines in the conventional areas of the economy, a host of jobs can and must be created in areas associated with the goal of sustainable development and environmental integrity. Parallel developments occur as well. Presidential commissions and national task forces are created by various nation-states to study and advise on how to provide new jobs that are interesting, challenging, and satisfying, as well as to look into the implications of shortened work weeks, longer vacations, retraining as part of every job, and paid leave for community service work. This accommodation to information technology is carried out for reasons of social survival and pragmatic self-interest.

Conclusion

The transition to an Information Society is not a simple computer hardware, software, or systems exercise. The issues we face are fundamental. They affect our economic, political, and social institutions. The changes associated with this transition will set the international agenda into the twenty-first century.

The relative newness of the computerization of society has caught all of us, including governments and private corporations, off guard. It was thought that by issuing a policy here, monitoring a development there, we could, in general, muddle through. But information technology is transformative. It challenges the way we measure and manage our economy. The authority of, and confidence in, existing institutions is being constantly eroded. A window of opportunity still exists, but it is closing fast.

Information technologies challenge both how governments are organized and the mandates of their individual departments. Most institutions, businesses, and government departments are in place to deal with yesterday's issues. Often structures and mandates refer to an earlier age of smokestacks, freight cars, and manufacturing—a time when things could be measured and policies and programs could be based on adding something more here and taking something else away there. However, information technology is changing our notions of work and working, the quality of time spent at home, and how we use our leisure time. In fact, it is blurring the boundaries between work, leisure, and education. Information technologies are changing our values, attitudes, and institutional arrangements.

We have reached the point of no return. There is no turning back. Like electricity, the internal combustion engine, and steam power, information technologies will be deployed and will transform our culture, society, and personal lives. No matter how hard we try to develop policies for information technology to evolve in an orderly way, we will have to face trade-offs. The transition to an information society brings vast disruption and personal and social dislocation. The productivity and efficiency inherent in an information technology that saves on labour, capital, and energy provide the major reason why society seems willing to go through such a wrenching set of changes.

In so many ways it is a magnificent technology quite capable of leading to the promised nirvana. But our society needs to take a flexible approach to the massive changes under way. Yesterday's ideologies and nostrums will not provide answers to the introduction of today's technologies. Sticking to yesterday's approaches may lead to system failures tomorrow. We need imagination. We need new theories, new perceptions, and institutional innovation across a broad front. The new constant is change. The solutions to this challenge must come from flexible and pragmatic approaches, and institutional innovation.

The Information Revolution will remove many of the constraints experienced by previous generations. Inherent in the technological advances is the potential for abundance, new and more interesting employment for some, added leisure for others, and longer and healthier lives for all. By adhering to old constraints and outdated ideologies, whether of the left or the right, we do a disservice to our fellow citizens and run the risk of missing out on the enormous benefits that these new tools, properly used, can bring with them.

Although the nature of work will surely change, it is possible to ensure that the transition to an information society will take place smoothly. For this to happen we must realistically face up to the choices and trade-offs that will confront us over the next decade. Only by looking at the costs and benefits and their social impacts can we be assured that the transformations will take place in a way that, to quote the English philosopher Jeremy Bentham, ensures "the greatest happiness of the greatest number."

Notes

1. This point was originally made in T.R. Ide, "The Left After the Crisis of Communism," paper presented to a conference organized by Fundacion Sistema, Madrid, Spain, Dec. 12-14, 1991.
2. An article in *The Toronto Star*, March 22, 1992, placed the number of middle managers and professionals who had lost their jobs in the United States and Canada during the past decade at more than a million. See also *The Globe and Mail*, March 4, 1992, p.A1.
3. Ivan Illich, *Tools for Conviviality* (New York: Harper & Row, 1973).
4. Gunter Friedrichs and Adam Schaff, eds., *Microelectronics and Society: For Better or for Worse* (Oxford: Pergamon Press, 1982).
5. Robert B. Reich, *The Work of Nations* (New York: Alfred A. Knopf, 1991).
6. John Ruskin, *Unto This Last, and Other Writings* (London: Penguin Books, 1985), p.199.
7. This point was originally made in Ide, "The Left After the Crisis," p.2.
8. Economic Council of Canada, *Good Jobs, Bad Jobs: Employment in the Service Economy* (Ottawa: Canadian Government Publishing Centre, 1990).
9. Ruskin, *Unto This Last*, p.222.
10. Richard M. Cyert and David C. Mowrey, "Technology, Employment and U.S. Competitiveness," *Scientific American*, 1989, p.54.
11. Organization for Economic Co-operation and Development (OECD), *The Diffusion of Advanced Telecommunication in Developing Countries* (Paris, 1991), p.33.
12. OECD, "Diffusing New Technologies: Microelectronics," *Science, Technology, Industry Review*, (Paris, 1989).

13. Chris de Bresson, *Understanding Technological Change* (Montreal: Black Rose Books, 1987), p.180.

14. Arthur J. Cordell, "Information Technology and the Commoditization of Content," in *Technology and Work in Canada*, ed. Scott Bennett, Canadian Studies series, Vol.8 (Toronto: Edwin Mellen Press, 1990).

15. Paul Wallich, "Going Through Proper Channels," *Scientific American*, March 1992, p.26.

16. W. Daniel Hills, "The Connection Machine," *Scientific American*, special issue, "Trends in Computing," 1988, p.24.

17. David Gelernter, "The Metamorphosis of Information Management," *Scientific American*, August 1989, p.69.

18. Andrew Topper, "Automating Software Development," *IEEE Spectrum*, November 1991, p.56.

19. Northrop Frye, "On Value Judgments," in *Criticism* (Madison, Wis.: University of Wisconsin Press, 1968), p.37.

20. Frederick Williams, *The New Telecommunications: Infrastructure for the Information Age* (New York: The Free Press, 1991).

21. Ibid., p.5.

22. Harlan Cleveland, "Governing in an Information Society," draft presentation to the Institute for Research on Public Policy, Ottawa, April 1992.

23. Canada Consulting Cresap, "Banking," report prepared for the Science Council of Canada, Toronto, 1991, p.94.

24. Ibid., p.95.

25. Hanley, Leonard, and Glossman, "Technology and Banking: Positioning for the 1990s," report prepared for Salomon Brothers, New York, 1991.

26. "Downward Mobility," *Business Week*, March 23, 1992.

27. Ros Easson, "Employment in Telecommunications," *CIRCIT*, Melbourne, Australia, April 1992.

28. *Wall Street Journal*, March 4, 1992, p.K16.

29. "Smart Programs Go to Work," *Business Week*, March 2, 1992, p.96.

30. Economic Council of Canada, *Good Jobs, Bad Jobs*, p.12.

31. Ibid., p.14.

32. Gary Burtless, ed., *A Future of Lousy Jobs: The Changing Structure of U.S. Wages* (Washington, D.C.: The Brookings Institute, 1990).

33. "Downward Mobility, *Business Week*, March 23, 1992, p.58.

34. Myron Magnet, "The Rich and the Poor," *Fortune*, June 6, 1988.

35. *The Toronto Star*, May 8, 1992, p.A1.

36. *The Globe and Mail*, May 12, 1992.

Afterword

Ursula M. Franklin*

In a series of lectures delivered just before his death in 1977, E.F. Schumacher examined the availability of what he called "good work" in a technological society. The famous theorist of economies of (rational) scale took as his point of departure Albert Camus's statement that "without work, all life goes rotten, yet, when work is soulless, life stifles and dies."

Schumacher pointed out that the spread of large-scale industrial and managerial technologies would inevitably bring both a sharp decrease in employment opportunities and a severe deterioration in the meaning and substance of the remaining work opportunities.

As I read *Shifting Time*, I felt compelled to look again at Schumacher's book *Good Work*. The present volume is an essential companion piece to Schumacher's analysis of the paradoxes inherent in technological societies.

Armine Yalnizyan, Ran Ide, and Arthur Cordell address the reality of the decline of the availability and meaning of work in Canada. This book speaks to the human condition in a world

*Ursula M. Franklin, C.C. FRSC, Professor Emerita, Massey College, University of Toronto, is an experimental physicist and the author of *The Real World of Technology*, adapted from her 1989 Massey Lectures.

where technological systems are characterized not by human-scale endeavour but by greater complexity, increasing capital cost, and accelerating violence. Because of these dominant trends, Schumacher urged a focus on appropriate technologies, on ways of work and production guided by his famous maxim "economics, as if people mattered." With talk of jobless recovery, this needs to be extended to "work, as if workers mattered."

After two decades of public policies that have paid no attention to the role of technology in the face of a clear deterioration of civil societies *and* the world's natural environment, governments in Canada and elsewhere are promising "job creation." Any job will apparently do. Never mind good work, steady work, meaningful work.

Yalnizyan, Ide, and Cordell—three writers of remarkably varied background—urge their readers to examine the new realities of work as well as the fate of those who may never find work, Canadians whose path to meaningful social contribution and economic independence is blocked by "technological progress." But how are citizens, both individually and collectively, to go about such an examination?

Two issues seem central to any public discourse. One is the social and political impact of technology. The other is the nature and scope of governance in a technological world. I have attempted to address the first, as have numerous other writers including Schumacher, Jacques Ellul, and Margaret Lowe Benston.

But it is much more difficult to find a meaningful focus for the second issue. What are the places and roles of government, of law, of regulation, of public policy in a technological world? Are they to make the world safe and profitable for technology and those who deploy it? If so, who gave their consent to this use of public policy, public funds, and public law enforcement?

Any constructive approach to work and the human condition

in a technological world will depend on a concerted effort by people to constrain and re-orient the uses of technology. At the same time we must insist on new priorities in governance, for it is essential on every level to govern and regulate as if people mattered.

Suggested Reading

Daniel Drache and Harry Glasbeek. *The Changing Workplace: Reshaping Canada's Industrial Relations System.* Toronto: Lorimer, 1992.

Ann Duffy and Norene Pupo. *Part-time Paradox: Connecting Gender, Work and Family.* Toronto: McClelland & Stewart, 1992.

Economic Council of Canada. *Good Jobs/Bad Jobs: Employment in the Service Economy.* Ottawa, 1990.

Ursula Franklin. *The Real World of Technology.* Toronto: Anansi, 1990.

Andre Gorz. *Farewell to the Working Class: An Essay in Post- industrial Socialism.* Boston: South End Press, 1980.

Morley Gunderson and Leon Muszynski with Jennifer Keck. *Women and Labour Market Poverty.* Ottawa: Canadian Advisory Council on the Status of Women, 1990.

Ivan Illich. *Tools for Conviviality.* New York: Harper Colophon, 1973.

Paul Kennedy. *Preparing for the Twenty-First Century.* New York: Random House, 1993.

Linda McQuaig. *The Wealthy Banker's Wife: The Assault on Equality in Canada.* Toronto: Penguin, 1993.

National Forum on Family Security. *Family Security in Insecure Times.* Ottawa, 1993.

Bruce O'Hara. *Working Harder Isn't Working: How We Can Save the Environment, the Economy, and Our Sanity by Working Less and Enjoying Life More.* Vancouver: New Star Books, 1993.

Juliet B. Schor. *The Overworked American: The Unexpected Decline of Leisure.* New York: Basic Books, 1991.

Guy Standing. "Alternative Routes to Labour Flexibility." in *Pathways to Industrialization and Regional Development,* ed. M. Storper and A.J. Scott. London: Routledge, 1992.